Praise for Murky wa[ter]

At last, a genuinely fresh [...] d
utilities are failing. Throu[gh ...] ;
this book tells a rich and [...]
pay for the huge investme[nt ...]
how it can be afforded wh[...], rather than forcing
poorer households to pay pr[oportio]nately more than the rich. Essential reading.
Aditya Chakrabortty, Senior Economics Commentator, Guardian

Murky water is a rigorously researched and well-argued book that cuts through the mire and offers clear and practicable solutions to the water crisis in the UK. It recognises that Westminster politicians will not tackle the crisis effectively unless pressured by us as citizens organised into a social movement for water reform.
Grace Blakeley, author *Vulture Capitalism*

To rebuild water management, nothing less is needed than a disruption from below countering the disruption from above by the financialised power structure that has taken possession of the sector. The book makes a convincing case for social movements as indispensable agents for restoring democratic control over the basic needs of economic and social life.
Wolfgang Streeck, author of *Taking Back Control?*

A meticulously researched tale of the greed, incompetence and regulatory failures that have brought the UK's privatised water system to its knees.
Gill Plimmer, Infrastructure Correspondent, *Financial Times*

Few issues provoke so much righteous anger in twenty-first-century Britain as the profiteering from and neglect of water. The Foundational Economy authors provide the ultimate audit of Britain's water crisis, recombining its financial, regulatory and material dimensions. The result is both an undeniable indictment of neoliberal failure and a

handbook for a different form of materialist political economy in relation to our most essential resource.
William Davies, author of *This is not Normal: The Collapse of Liberal Britain*

Too often the economic and the social are treated as separate domains and dealt with in separate analyses. *Murky water* is a book that innovatively brings together what's gone wrong economically in privatised water and shows how we can fix it socially if we mobilise for change.
Hilary Cottam, author of *The Work We Need*

A singular achievement that brings a radically new approach to bear on the provision of basic services. This fascinating study disposes of the mythology that our water and sewage systems are Victorian leftovers, showing that they essentially date from the long post-war boom. The failure to invest since is not just the result of extractivism and compromised regulation but of the very system of charging for water, which weighs too heavily on the poor and too lightly on the rich. If we are to live better we need to attend to the necessities of life in new ways suggested by this vitally important book.
David Edgerton, author of *The Rise and Fall of the British Nation*

A serious attempt to grapple with the issues not only of ownership but of accountability for performance and equity in terms of the price of water. Too often on the left there is an automatic, almost kneejerk assumption that public ownership is desirable without a serious consideration of the difficult issues of effective governance and accountability, the source of the considerable required investment and the pricing of water for users.
Andrew Davies, formerly Minister for Economic Development, Welsh Government

Combines forensic analysis with easy readability to tell the full story of how our water system has come to be in such an appalling state. The accounts of policy naivety, political inaction and corporate greed will make your eyes water. Thankfully the authors also provide an

insightful vision of what needs to change and how. This book will both enrage and inspire.
Kate Bayliss, SOAS, University of London

Keir Starmer's government is doing everything in its power to avoid the demand of a large majority: that water in England must be taken back into public ownership. This book explains why and sets out how things will get worse if we don't break the model of financialisaton. The prognosis is terrifying, but it is not paralysing. The path the book offers is a means of taking back control of our water supply and funding it properly with progressive charging based on household income. This is much more than a guide to the ruins. It also offers us a way out of this mess. The authors have lit a fuse that can ignite a new movement for water justice.
David Whyte, author of *Ecocide*

Murky water

Manchester University Press

The Manchester Capitalism book series

Manchester Capitalism is a series of short books that follow the trail of money and power across the systems of financialised capitalism. The books make powerful interventions about who gets what and why, with rigorous arguments that are accessible for the concerned citizen. They go beyond simple critiques of neoliberalism and its satellite knowledges to re-frame our problems and offer solutions about what is to be done.

Manchester was the city of both Engels and Free Trade where the twin philosophies of collectivism and free market liberalism were elaborated. It is now the home of this venture in radical thinking that primarily aims to challenge self-serving elites. We see the provincial radicalism rooted here as the ideal place from which to cast a cold light on the big issues of economic renewal, financial reform and political mobilisation.

Books in the series so far have covered diverse but related issues. How technocratic economic thinking narrows the field of the visible while popular myths about the economy spread confusion. How private finance is part of the extractive problem not the solution for development in the Global South and infrastructural needs in the UK. How politics disempowers social housing tenants and empowers reckless elites. How foundational thinking about economy and society reasserts the importance of the infrastructure of everyday life and the priority of renewal.

General editors: Julie Froud and Karel Williams

To buy or to find out more about the books currently available in this series, please go to: https://manchesteruniversitypress.co.uk/series/manchester-capitalism/

Murky water

Challenging an unsustainable system

Luca Calafati, Julie Froud,
Colin Haslam, Sukhdev Johal and
Karel Williams

MANCHESTER UNIVERSITY PRESS

Copyright © Luca Calafati, Julie Froud, Colin Haslam, Sukhdev Johal and Karel Williams 2025

The right of Luca Calafati, Julie Froud, Colin Haslam, Sukhdev Johal and Karel Williams to be identified as the authors of this work has been asserted in accordance with the Copyright, Designs and Patents Act 1988.

Published by Manchester University Press
Oxford Road, Manchester, M13 9PL

www.manchesteruniversitypress.co.uk

British Library Cataloguing-in-Publication Data
A catalogue record for this book is available from the British Library

ISBN   978 1 5261 8869 4   hardback
ISBN   978 1 5261 8870 0   paperback

First published 2025

The publisher has no responsibility for the persistence or accuracy of URLs for any external or third-party internet websites referred to in this book, and does not guarantee that any content on such websites is, or will remain, accurate or appropriate.

EU authorised representative for GPSR:
Easy Access System Europe, Mustamäe tee 50, 10621 Tallinn, Estonia
gpsr.requests@easproject.com

Typeset
by New Best-set Typesetters Ltd

Contents

List of exhibits vii
Abbreviations xi

Introduction **1**
    Three stories: extraction, service failure and environmental challenge 1
    Three foundational threads: numbers, narrative and history 7
    The mess we are in and what to do 24

1 From sanitation to water management **30**
    1.1 Not a simple business 32
    1.2 Sanitation's technical and political history 40
    1.3 Water management under climate change 50

2 The business model problem **63**
    2.1 The business model problem 66
    2.2 The revenue constraint 81
    2.3 How ownership matters 89
    Appendix: business users and competition 101

3 Much more investment required **105**
    3.1 Decades of patch and mend 107
    3.2 Water UK's fantasy plan 116
    3.3 Dawning realism about climate adaptation 126

Contents

4 From water poverty to water justice — **139**
   4.1 Charging through the lens of water poverty — 141
   4.2 Regressive charging and the limits of social tariffs — 150
   4.3 The justice case for flat-rate and progressive charging — 164

5 Failure of political and regulatory control — **178**
   5.1 Political control as issue management — 180
   5.2 Ofwat: economic regulation of business — 191
   5.3 The Environment Agency: sponsored regulation — 203
   Appendix: avoiding abstraction — 213

6 Towards foundational water management — **221**
   6.1 Foundational water management and the right moves — 224
   6.2 Zombie companies and projectification — 238
   6.3 Power configurations and policy pathways — 251
   6.4 Towards a social movement that challenges power — 261

Notes — 274
Index — 326

Exhibits

The data used to create the exhibits is available at https://foundationaleconomyresearch.com/index.php/murky-water-stats

1.1	Change in global CO_2e emissions and world GDP (inflation-adjusted) since 1990	54
1.2	Annual variation in temperature relative to the long-run trend in England, 1884–2023	58
1.3	Annual variation in precipitation relative to the long-run trend in England, 1836–2023	58
1.4	Seasonal linear trends in precipitation across England, 1836–2023	60
2.1	Post-tax income, dividends and change in reserves of water companies in England and Wales (inflation-adjusted), 1989–2023	72
2.2	Analysis of asset intensity in various utilities and other businesses, 2012–23: (a) the value of physical assets used to generate each £1 of sales revenues; (b) the value of annual	

List of exhibits

	capital expenditure on physical assets per £1 of sales revenues	75
2.3	The distribution of operating cashflow for capital expenditures, interest and dividends among four utilities since their privatisation or creation	80
2.4	Average water company revenue per dwelling in England and Wales, 1990–2023	86
2.5	Total water company sales revenue (inflation-adjusted), categorised by price review period, 1989–2023	87
2.6	Changes in ownership of water and sewerage companies since their privatisation	91
2.7	Claims made on sales revenues and operating cashflow in English and Welsh water companies, by ownership type: (a) capital expenditure, dividends and interest as a share of sales revenues; (b) capital expenditure, dividends and interest as a share of operating cashflow	96
2.8	Thames Water equity and debt, 1989–2023: (a) equity and debt as a share of total capital; (b) value of shareholder equity and debt	99
3.1	Capital expenditure and operating expenditure of water companies across price review periods, 1989/90–2023/24	111
3.2	The proportion of newly installed or renovated water and sewer mains since 2001	112
3.3	Water UK's expected annual frequency of anticipated spills from storm overflows in England, 2025–50	120

List of exhibits

3.4	Current and anticipated seasonal average temperatures and precipitation levels, 1981–2000 and 2080–99	129
3.5	UK Climate Change Projections (UKCP18) for England and Wales: average summer temperature and changes in precipitation	131
3.6	Daily average water consumption and leakage per person in England, 2003/04–2022/23	134
4.1	Household annual average water bill by region, 2023	149
4.2	Household average water bills and the revenue raised in England and Wales in 2023, by decile (total revenue £9.505 billion)	153
4.3	Average water bill per household and per person, England and Wales, 2023	155
4.4	Simulated doubling of the 2023 revenue raised in England and Wales and the amount paid by each decile (total revenue £19.011 billion)	158
4.5	Simulated household water bills in England and Wales based on doubling the amount raised, with a 50% discounted social tariff for deciles D1 to D3	163
4.6	Simulated 2023 water revenue of £9.5 billion, raised using a flat-rate charge of 0.86% of household total expenditure, by decile	166
4.7	Simulated doubling of the 2023 water revenue to £19 billion, raised using a	

List of exhibits

	flat-rate charge of 1.72% of household total expenditure, by decile	167
4.8	Analysis of household expenditure on water, contrasting the current system in 2023 with flat-rate and progressive simulations	175
5.1	Corporate structure of Kemble Water Holdings, owner of Thames Water, 2022/23	200
6.1	English and Welsh water companies' debt and debt servicing, 1990–2022	242
6.2	Strategic Infrastructure Procurement Route, Direct Procurement for Customers and in-house major projects agreed by Ofwat (in 2022/23 prices)	246
6.3	Vintage of reservoirs in England and Wales, showing the percentage of the 2025 reservoir capacity added in each time period, 1652–2025	256

Abbreviations

BBC	British Broadcasting Corporation
BT	BT Group PLC (formerly British Telecom)
CCW	Consumer Council for Water
CMA	Competition and Markets Authority
CPI	consumer price index
Defra	Department for Environment, Food and Rural Affairs
DPC	Direct Procurement for Customers
IPCC	Intergovernmental Panel on Climate Change
KPI	key performance indicator
LCF	Living Costs and Food Survey
NAO	National Audit Office
NIC	National Infrastructure Commission (replaced in April 2025 by the National Infrastructure and Service Transformation Authority)
OEP	Office for Environmental Protection
Ofwat	Water Services Regulation Authority
ONS	Office for National Statistics
PFI	Private Finance Initiative
PLC	public limited company

Abbreviations

PR	price review
RPI	retail price index
SIPR	Strategic Infrastructure Procurement Route
SSWAN	Sustainable Solutions for Water and Nature
WASP	Windrush Against Sewage Pollution
WRMP	Water Resources Management Plans

Introduction

Three stories: extraction, service failure and environmental challenge

Murky (adjective)
1 dark and dirty or difficult to see through
2 used to describe a situation that is complicated and unpleasant, and about which many facts are not clear[1]

The operations of the privatised English and Welsh water and sewerage systems are murky in both senses of the word. They are murky because so much is obscure and unclear, and murkier still because so much of how these systems operate is complicated and dubious. In this book we aim to cut through the murk and reveal what is defective. We begin by considering the stories that represent current public knowledge about water and sewerage. The public understands what has gone wrong in privatised water and sewerage in England and Wales through three separate stories: a financial extraction story, a failure of service story and an environmental challenge story. The brief descriptions below show how two of these stories give us affective, satisfying problem definitions where we can 'hiss the villain', but do not give us much grip on solutions. We turn to the concept of the foundational economy to open up a different kind of analysis that might point to such solutions.

The financial extraction story is a moral story about profits. The profits of privatised water companies were distributed in dividends to shareholders while executives took high rewards, and investment was financed by issuing more than £60 billion of debt. Chapter 2.1 explains how this story was originated by researchers Kate Bayliss and David Hall and was accepted across the political spectrum after Michael Gove as Secretary of State at the Department for the Environment, Food and Rural Affairs (Defra) made a speech excoriating the companies in 2018. The story was subsequently circulated by mainstream media with the addition of minor refinements. For example, Nils Pratley in *The Guardian* adds 'political and regulatory desire for lower bills' to 'a toxic mix of dividend extraction, debt accumulation and financial engineering'.[2] This is above all a moral story, as we see from the tone of the Compass 2025 report on the privatised water industry.

> there was always more than enough money. Instead of investing in the water network, the money from bills was used to line the pockets of executives, shareholders, and bosses. England's privatisation of water, perhaps unsurprisingly, has been an exercise in profit extraction – not providing a service. The regulators have been defanged to the point of inefficacy and the companies are increasingly uninvestable.[3]

The failure of service story is a disgusting story about sewage disposal. This centres on irresponsible companies dumping raw sewage into water courses through 15,000 storm overflows. These overflows should discharge only in heavy rain but are incontinent, and the discharges include 'dry spills' when there has been no

Introduction

rain. This is a disgusting story because it is driven by cultural and ecological values and by indignation about the gross irresponsibility of the companies. The story originated in civil society non-governmental organisations (NGOs) such as Windrush against Sewage Pollution, which used pioneering citizen science to show that, from 2010 onwards, the Environment Agency was seriously under-recording sewage spills.[4] Campaigning by civil society raised the profile of the problem, which was then taken up in various media investigative reports. For example, the British Broadcasting Corporation (BBC) in 2024 reported that United Utilities in north-west England had pumped 140 million litres of untreated sewage into Windermere between 2021 and 2023. Corporate irresponsibility was highlighted through a company whistleblower's account of a cover-up:

> it would have flagged up on flow and spill reports. I'm not surprised we haven't reported it. We work on a risk management basis and we'll have judged the risk of the Environment Agency finding out and understanding the permit breach would be minimal as they are under-resourced and incompetent.[5]

The environmental challenge story is a broader, more cerebral story. This story is about the many sources of nature-depleting pollution, which include agricultural and road runoffs as well as sewage dumps. It also looks towards tomorrow's challenge of climate change bringing new and widespread problems of both water excess and water deficiency. This story has been driven by environmental action NGOs such as River Action, which was initially concerned with the degradation of the river Wye through runoff from intensive poultry farming. As Chapter 3.3 observes, the environmental challenge had by 2023 crossed over into official reports by the Environment Agency and Defra, backed up by others from the National Infrastructure Commission on the investment required and from the Office for Environmental Protection on the failure to meet environmental targets. The scale and nature of these environmental challenges makes this a cerebral story whose tone is captured in the Rivers Trust evidence on the water sector for a 2024 parliamentary committee.

> Ultimately, we need to stop seeing the water sector as a discrete unit and start thinking of water companies as a key partner in a collaborative approach to addressing our huge environmental challenges, with money being invested where it can have the greatest impact and deliver the most benefits to climate change resilience and mitigation, nature recovery, water resource resilience and flood risk.[6]

The stories remain distinct even when they are combined in sequence, as when financial extraction is given as cause of failure of service. Separately or in combination, these stories circulate widely and powerfully because they work affectively through two classic narrative tropes: 'we name the guilty men' and 'arrow indicates defective part'. Thus, the public has the shareholders and executives as the guilty men and the storm overflow as the defective part. This is both satisfying and fits the 'shocking facts' highlighted in the stories. There *has* been an irresponsible distribution of dividends by water companies at the expense of financial reserves, as Chapter 2.1 confirms, just as raw sewage was spilled into waterways for 3.6 million hours in 2024.[7] The third story of broader environmental challenge is slowly gaining profile, but this story's circulation is limited by the way its seriousness is not vindicated by everyday experience nor consensus endorsement. As Chapter 1.3 demonstrates, the United Kingdom has already had an average 1.5 degree temperature increase compared to pre-industrial levels, with wetter winters and drier summers,

Introduction

but this is obscured by large year-on-year variability. Meanwhile, debate about adaptation to climate change is crowded out by rhetorical hostility to what Conservative leader Kemi Badenoch terms the 'fantasy politics' of net zero mitigation.[8]

The popular stories encourage 'hiss the villain' problem definitions, but none of the stories comes with a fixed line on what to do by way of solution. Ideas about 'what to do' are added according to the left, right or centrist political position of the storyteller. On the centre and the right, the aim is to change the behaviour of the polluters (whether water companies or farmers) and reform regulation so as to make it more effective. Thus in 2024, the Labour government introduced the Water Measures Bill to deal with corrigible misbehaviour by water companies and appointed the Cunliffe Commission, whose brief included regulatory reform.[9] The left adds an emphasis on accountability and redressing a 'democratic deficit'. Thus the 2025 Compass Report complains that 'the group of people with a seat at the table in the governance of English water does not include anyone with a legitimate democratic mandate to serve our people and our environment'.[10] This leads to the issue of public ownership, which is proposed, as by Labour at the 2019 election or the Greens at the 2024 election, with the aim of ending profit extraction and introducing democratic accountability.

However, public ownership is not on the official agenda, creating a longstanding divide between the majority of citizens who want public ownership and the centrist political classes in Westminster and Whitehall who will not consider it.

- A majority in the UK did not want water privatisation and an increasing number are not reconciled to it. The policy was so unpopular in the 1980s that the Tories pulled a privatisation proposal from their 1987 election manifesto. As Chapter 2.1 notes, opinion polls show that more than 80% now favour public ownership. This is a clear preference, but it leaves many issues about what public ownership means in practice unresolved.

- Westminster and Whitehall will not countenance public ownership. Although private ownership is discredited, as Chapter 5.1 observes, neither Labour, the Conservatives nor the Liberal Democrats included public ownership in their 2024 manifestos. The new Labour government appointed the Cunliffe Commission in 2024 with a brief to consider many reform options, but not public ownership. Moreover, as Thames Water stumbled into distress borrowing in 2025, Defra and the Water Services Regulation Authority (Ofwat) did not invoke the special administration regime because that could be construed as temporary nationalisation.

Some NGOs, exercised by failure of service or environmental issues, treat public ownership as an awkward issue that is best avoided. Support for public ownership may limit their dialogue with government and companies, and make it more difficult to obtain broad-based support. As Chapter 6.4 observes, public ownership did not figure in the 2023 End Sewage Pollution Manifesto which preceded the one-day protest of the 2024 March for Clean Water. The march was supported by more than a hundred organisations, including Water UK, the trade association of the privatised water companies. Thus, the problems of water may be grave, but radical change is often not on the agenda.

These three stories give us accounts of how the industry has failed and is still failing, which lead to disagreements about how it could be fixed. They also leave many questions unanswered,

Introduction

and so water remains murky. We need to go further, and that is what this book is about.

Three foundational threads: numbers, narrative and history

We aim to go further in both understanding problems and outlining actions through foundational analysis. The foundational economy approach focuses on the provision of everyday essentials such as food, housing, health services, utilities and transport within planetary limits. It is a developing way of thinking both descriptively and analytically about what matters in a changing world, and thinking normatively about how to change it for the better. The development works by carrying over old elements and embedding them in new analysis of the changing conjuncture. The original descriptive phase, up to and including the 2018 foundational economy book,[11] was a rejoinder to industrial policy's concerns with exciting sectors such as high tech. What matters more for ordinary citizens is the foundational economy, which employs around 40% of the workforce and provides households with basic goods and services such as healthcare, education, water, energy and food. Foundational thinking has distinguished between material systems such as energy and water, and providential systems such as health and care, emphasising the importance of understanding the distinctiveness of each.

Foundational economy took an analytic turn in the 2023 book *When Nothing Works*, which engaged with the 'cost of living crisis' in a world of nature and climate emergency.[12] This added an emphasis on household liveability, introducing the three-pillar concept of liveability which reflects residual income, essential

Murky water

services and social infrastructure. This book on water is the first attempt at an extended analysis of how a major foundational system works in terms of both the supply of services over the long term and the households who pay for this. The analysis covers not only what has gone wrong in a system whose function is the collective provision of clean water and the disposal of wastewater, but also where intervention should be focused.

In starting to think about how the water system does work, we start with a normative foundational concept of how it should work. As Chapters 1.1 and 6.4 observe, water is often discussed in generic terms using the language of human rights or the commons. This language has powerful resonances, but if the aim is affordable access to, and control over, a foundational system that matters to households and the planet, foundational water management needs to have a specific focus. From a foundational perspective, there are five desiderata that reflect the liveability and sustainability aspirations for the water system.

1. *Technically capable*, that is, delivering potable quality water and disposing of wastewater in the face of both water shortage and excess as a result of climate change.
2. *Socially and spatially inclusive*, that is, capable of servicing all households to a high standard in all regions.
3. *Financially accessible*, that is, with costs charged equitably so as not to overburden low-income households or undercharge high-income households.
4. *Environmentally responsible*, that is, with potable water sourced sustainably and waste disposed of in ways that are not environmentally damaging.

Introduction

> 5 *Democratic*, that is, with an informed and organised public influencing decision making and the overall dynamic and tendency of the system.

As pragmatists, we recognise that a foundational system which meets all these criteria is a utopian ideal. The brief technical and political history of sanitation in Chapter 1.2 shows that the water system has never met more than three of the five desiderata. Historically, the English and Welsh system was undemocratically designed by elites at the expense of communities that got in the way, and it has until very recently treated the environment as a sump for abstracted water supply and a sink for out-of-sight-and-out-of-mind waste disposal. If systems fall short of meeting these criteria, the first practical question about any system is about tendency and direction: is the water system moving towards or away from the foundational ideal?

The dispiriting answer is that the current English and Welsh system is moving away from the foundational ideal. Most obviously, there are financial accessibility and affordability issues. After twenty years when household water bills did not increase in real terms, we now have the prospect of rising bills stretching into the future as far as we can see. The media report that the first of these price rises will produce a year-on-year increase of £123 in the average household water bill in 2025–26 across England and Wales; Southern Water consumers will see a 47% increase to an average bill of £703.[13] The immediate environmental issue is that, as Chapter 3.2 argues, planned water company investments will not quickly solve storm overflow problems. Taking a broader perspective, Chapter 3.3 notes that

Murky water

Defra – as the responsible government department – does now recognise climate change issues, but also argues that there is no credible strategy for dealing with the resulting dual problems of water excess and shortage. As Chapter 6.4 explains, everything will be made worse by projectification, that is, relying on private finance consortia to design, build and operate new reservoirs, sewerage treatment plants and water transfer schemes which are now necessary after twenty-five years of inadequate physical investment.

How, then, do we identify the points of intervention and the levers that could change the direction of the water system and set it on a more foundational pathway? That requires an analysis of the water system which, as with our previous work, is empirically resourceful and conceptually minimalist. We lean heavily on forensic research into statistical sources both in corporate accounts and official statistics, because we do not think there is any set of concepts or method which is adequate to an object as complex as the water system. Instead, the book develops a narrative and numbers analysis which can be traced back to our work on financialisation from nearly twenty years ago.[14] This analysis works by developing and relating three threads of argument and evidence.

1 The first thread is about numbers in the financialised operating system of privatised water, as we track how money moves from households to water companies, dealing with hard-to-reconcile claims on the companies' cashflow.

Introduction

> 2 The second thread is about the narratives produced by actors such as Defra and Water UK, for whom narratives are about distracting the public when what we need are narratives that focus effective collaboration and action.
>
> 3 The third thread offers a historical perspective which is about how and why top-down power relations have opened and closed possibilities of action in the changing Overton window about what is thinkable and doable.

Tracking these threads takes the reader on a journey that features Victorian sewers and post-war dams, company business models and privately financed major projects, extractive finance and the power of political, technical and financial elites. The result is not analysis but a non-linear narrative which works through time shifts, converging storylines and multiple perspectives. This book is less like traditional social science and more like a classic art movie such as Gonzalez Inarritu's *Amores Perros*, which – like our book – works through devices designed to seed multiplicity. Chapters are divided into thematic sections which move through distinct themes, as in Chapter 1, where section 2 explores the difference between the post-1850 technical and political histories of sanitation and section 3 focuses on climate change since 1880. Depth is added in the way that many sections have a double object. Thus Chapter 3.2 is about the inadequacy of water industry investment plans and about how trade narratives deceive the public; Chapter 5.3 on the Environment Agency explores the failures of environmental regulation and how the Agency provides political cover for ministers. There are two appendices which add important elements and show how the threads are interwoven.

The threads of argument and evidence are developed in five chapters of analysis. The sixth and final chapter presents our 'what to do' argument. Many of the chapters rely on statistical evidence which is generally presented in charts. For each exhibit, the supporting numbers are available on the Foundational Economy Research website. The statistical material is extensive because wherever possible we include time series since privatisation in 1989. This adds complication but gives invaluable perspective. The analysis covering 'water companies' focuses on the ten regional water and sewerage companies which were created in 1989, but does not cover the very much smaller water-only companies. Water companies and water industry are used throughout the text as shorthand for these ten large water and sewerage companies, which cover over 80% of customers in England and Wales.

So, what do we find in the first five chapters by exploring each thread and what are the implications for action?

How money moves: the operating thread

Financial and physical numbers provide rich but tricky material. Critics of financialisation focus on greedy and irresponsible behaviours by company managers and investors allocating cash between three demands: the expenses incurred in operating the water and sewerage systems; physical investment to maintain or enhance the system; and interest and dividend payments on finance capital. Understanding the business model characteristics

Introduction

of an asset-heavy infrastructure does not excuse extractive corporate behaviour, but it is clear that corporate decisions about allocating cash to these three purposes are made in a financial and operational context that can be enabling or constraining. The upper financial limit is set by the flow of revenue from the paying customers, which are mainly households in the case of water. The physical limits on reallocation are set by the activity characteristics of a specific production and distribution process, which shapes the very different possibilities for reducing operating expenses (such as employment costs) and has very different requirements for physical investment. Thus, companies allocate cash and distribute profits in a financial superstructure which sits on top of a revenue and activity base. The financial extraction story outlined earlier in this chapter is an account of the financial superstructure, which is true enough, but abstracts from the base activity and revenue constraints in the system. Chapters 2 to 4 show that water has a business model problem relative to recovering its costs, regardless of public or private ownership, given the activity constraint of asset intensity and the revenue constraint of the existing household charging system.

The physical activity constraint is analysed in Chapter 2.1. Water is (with railways and the electricity grid) one of three asset-intensive utilities with a voracious demand for physical investment. Thus, in the 2010s, the water companies physically invested 40 pence from every £1 of sales revenue. None of the asset-heavy utilities have enough revenue to cover day-to-day operations (wages and purchases), physical investment and a return on finance capital. As Chapter 2.1 notes, the problem is compounded in water by the very limited scope for cost reduction by reducing employment. Under different ownerships, managers in all three asset-intensive utilities have managed their activity constraint in the same way. First, they have mainly limited physical investment to patch-and-mend replacement, and effectively stopped large-scale enhancement investment to upgrade systems and increase capacity. Second, they have taken out loans so that all three of the asset-intensive utilities have more than £55 billion of debt (or debt and equity from rights issue, in the case of the electricity grid). In the 2020s the UK does not have an electric rail network, nor a dense network of fast chargers for electric vehicles, nor adequate sewage treatment capacity.

All the UK's asset-intensive utilities have since privatisation moved on a trajectory towards accumulating problems related to service failure and financial unsustainability with a wrecked balance sheet. In the case of water, the problem has been made worse by the debt-based financial engineering of irresponsible owners, and Chapter 2.3 discusses how ownership matters. It shows how Macquarie deliberately loaded Thames Water with debt between 2006 and 2017, so that it could secure 12% returns on its equity stake when the allowed return in the regulated asset base was around 4.4%.[15] If water companies were operated on a not-for-profit basis and had no shareholders, that would exclude predatory financial engineers, but it would not solve the activity constraint problem. Chapter 2.3 explains how and why the bond-financed, not-for-profit water company Dŵr Cymru could not invest more or reduce bills; on that basis it is not surprising that its environmental record in terms of sewage discharges is no better than the English companies.[16] And this should be sobering for our readers, because Labour's 2018 public ownership plan was for public, bond-financed, not-for-profit water companies. Free capital in the form of Treasury grant-in-aid would help manage water's problems because they would not

Introduction

require interest payments. But remember that the public water authorities were starved of Treasury funding in the 1980s, and capital grants for water are vanishingly unlikely in the 2020s. In any case, Treasury capital grants might be free to the water companies, but households would still pay in the form of taxes rather than water bills.

The activity constraint is insoluble (regardless of ownership) because there is not enough revenue in the system. Households contribute around 75% of water company revenue and the constraint arises from the system of household charging, which is analysed in Chapter 4. Charging is complicated because it is based on meters – both dumb and smart – and rateable values. But the overall logic of charging is simply regressive. As Chapter 4.2 explains, poor households at the bottom of the distribution pay relatively more than affluent households at the top of the distribution. Poor households typically have fewer people, so the poor pay more absolutely for water on a per-person basis. In the 2000s and 2010s there was no effective national social tariff scheme which would have rebated the bills of the poor. So, for twenty years Ofwat and the water companies became co-conspirators in low bills, which set a hard revenue constraint. Chapter 4.3 shows that the belated introduction of a broader and national social tariff will address regressivity but at the expense of creating new problems relating to take-up and cliff edges in the middle of the income distribution.

These ideas about revenue constraint help move us towards action. As Chapter 5.1 shows, many policy proposals – from Labour's 2018 proposal for public ownership to the Liberal Democrats 2024 profits tax – misdiagnose the water's industry's problem. They assume that the problem is diversion of revenue to greedy investors and this assumption then licenses the misconception that not-for-profit private or public ownership would solve water's problems. Against this, the underlying problem is deficiency of revenue, and the diversion of revenue to dividends makes this even worse. We can also add that the charging system is usually understood through the lens of water poverty, especially when water bills are now inexorably rising, as they are set to do from 2025 to 2030. As Chapters 4.1 and 4.2 show, this encourages ideas about extending social tariffs and offering poor households bill rebates whose cost will be charged to those households paying full bills. If we shift the paradigm from water poverty to water justice, then progressive charging for water – according to household income – is the key technical change which unlocks the possibility of reform by lifting the revenue constraint. As Chapter 4.3 demonstrates, progressive charging works not by narrowly 'taxing the rich' but by ensuring that two-income professional and managerial couples in deciles 7–10 pay their fair share for a collective and essential infrastructure.

Our counterfactual expenditure simulations in Chapter 4.3 show that flat-rate or progressive charging according to household income would lift the revenue constraint and make it possible to raise twice as much revenue in an equitable way, to fund much more physical investment, and to service new borrowing. With this point made, the role of public ownership becomes imperative. A water industry with a much larger revenue pot would be much more attractive for predatory investors unless they are blocked. Public (or at least not-for-profit) ownership is the prophylactic which can block more financial engineering. Change of ownership is necessary as a corollary to a reform of charging, but public ownership without charging reform is likely to achieve little and would most likely serve to discredit the whole idea of public ownership of utilities.

How narratives distract and focus: the framing thread

Narratives have a double function in the present conjuncture. They are essential for focused action because effective reform that addresses system problems needs both a clear direction and points of leverage. Equally, narratives are useful distractions produced by politicians and corporates who are not solving our problems but managing their own issues by tackling symptoms, with the aim of getting through short-term difficulties. The failure of service and the environmental challenge stories establish that water has intractable problems which both require focused action and encourage politicians and corporates to try to buy time through distraction. This means that clearing these proliferating and distracting narratives out of the way is an important task, before considering the limits of what can be achieved through public indignation about storm overflows and failure of service. All this finally highlights the crucial importance of a clear counternarrative which, in the case of water, should give a tight problem focus on water management in a period of climate change (which of course connects with other issues around security of supply, water justice, and river and coastal water quality).

Corporates and trade associations have learned to defend their interests with trade narratives, which usually work by selectively citing all the good things they have done.[17] This listing of benefits is not possible in an industry discredited by the failure of service story, so the trade narrative in water has to be about promises to do better. For example, Water UK, the trade association of the private water companies, has an extend-and-pretend investment plan for dealing with storm overflows. Chapter 3.2 analyses Water UK's *Storm Overflows Plan for England*, which creates the impression that the water companies are purposively investing to make good the industry's legacy problems in relation to storm overflows. But this is an extend plan because it pushes the attainment of the overflow reduction targets into the distant future in 2050, and then backloads the action required to achieve these targets into later years. And this is a pretend plan because the exacting long-term targets for reducing the number of spills are not what they appear to be. On closer examination, the benchmark year is chosen to flatter performance, and the targets come with undisclosed assumptions and explicit qualifications that effectively act as a get-out clause.

Political narratives are equally distracting, especially government narratives, which can be performative. In the 1990s New Labour perfected a model of responding to public concerns (articulated in focus groups and opinion polls) with performative reassurances through announcements of action. This is the model for Defra and Steve Reed in the 2024 government, working to reassure the public that something is being done (preferably without upsetting anybody). This was certainly the case with the Water Special Measures Bill introduced by Labour after the 2024 election. Chapter 5.1 explains that this focused on issues such as customer compensation for service failures and 'fat cat' executive salaries. Both of these issues are peripheral to serious reform. The Bill appeared tough but contained no provisions that senior executives could not easily live with: they could be sent to jail, but only if they obstructed an investigation, and

Introduction

their bonuses were restricted but not their basic salaries. These initiatives signalled the desire for improvement, but not its delivery insofar as they focus on symptoms not causes.

These trade and political narratives almost certainly have little impact on the broader public because those concerned by the failure of service story are in a state of indignation fed by media coverage, including reports of a record 3.6 million hours of sewage spills in 2024.[18] However, the history of sanitation in Chapter 1.2 shows that emotionally charged public indignation is not enough. Public disgust about sewage disposal moves things on, but in an uncontrolled way, so that the policy response often delivers the next instalment of the problem. Public disgust about storm overflows in the 2020s echoes earlier episodes of public disgust in the 1850s, 1880s and 1980s. The 1850s 'great stink' in London produced Joseph Bazalgette's intercept sewers, which shifted the discharges further down the Thames; and in the 1880s an outcry about mass drownings in untreated sewage discharged into the Thames (after the *Princess Alice* sinking) produced one hundred years of sewage sludge dumping at sea, including off the Essex coast. The danger in the 2020s is that the current disgust about the discharge of untreated sewage from storm overflows will only produce end-of-pipe steel and concrete solutions at the treatment works, which will expand sewage treatment capacity and construct holding tanks. In a combined sewer system, these types of measures are futile without a range of nature-based and other solutions to slow and retain water flow as beginning-of-pipe solutions, both to mitigate urban surface flooding and to reduce hydraulic flow with intense rain in a changing climate. In the appendix to Chapter 5 we also explore the issue of abstraction, which is a lower-profile crisis by comparison with storm overflows, but is absolutely critical in relation to both water supply and ecological quality challenges.

Setting distractions aside helps move us towards action through a clear counternarrative that gives a tight problem focus. It is crucially important to recognise that we have a new problem regarding water management and long-run adaptation to climate change. The old problem of sanitation for public health remains real enough, but is now enclosed in this wider problem of water management. The challenge of water management under climate change is analysed in Chapter 1.3 and Chapter 3.3, which explain how it is hard to hold the focus on water management and long-run adaptation to climate change. The political debate about the avoidable costs of meeting net zero targets tends to crowd out discussion of the unavoidable costs of adapting to climate change. The changes that have already happened are masked by year-on-year variations in temperature and rainfall. Official reports have focused on forecasts up to 2050 and exclude the more alarming likely changes by 2100, including up to 4 degrees of warming and more than one-third less summer rain. The magnitude of the adaptation problem was finally recognised in reports by Defra and the Environment Agency in 2023 and 2024. However, an adaptation problem of this scale requires catchment, regional and national planning of action by multiple stakeholders, including landowners and local authorities. This is unlikely to work through voluntary partnerships, and muscular planning remains officially unthinkable.

How power works: the closure thread

Taking a historical perspective is helpful because the location of power is easily taken for granted in any one period, and

Introduction

power relations are most easily exposed by considering their operation over time. If history rhymes but does not repeat itself, identifying recurrent power relations focuses attention on what we need to change, just as contrasting different outcomes over time helps us understand the power relations which can open up or close down policy options and determine pathways. Regrettably, history does not figure in this way in the stories about water, which present history as an arc of deterioration from privatisation in 1989 in the extraction story, or from the 1880s reference point in the climate change story. Looking backwards at power raises complexities which we can cut through by focusing on reservoirs, whose history is revealing of recurrent power relations and different outcomes. Like other infrastructures, water is about power relations. Its political processes always involve decision making by elites in representative bodies, which operate in a quasi-oligarchic way. And yet outcomes are very different in successive periods, and that is what has to be explained if we are to understand the obstacles we now face and find the levers of change.

In terms of recurrence, the political history of reservoirs is of top-down power relations through upland land grabs and clearances from the 1880s to the 1960s. In the north of England and midlands from the 1880s to the 1960s the major conurbations – Liverpool, Manchester and Birmingham – sorted out their water supply problems by constructing reservoirs. The topography of the urban midlands and the north-west of England was such that large reservoirs were most easily constructed by flooding valleys in the Lake District and mid-Wales, with long distance gravity-assisted transfer to the urban users. The big municipalities used Acts of Parliament to flood valleys without consultation or inquiry. As Chapter 1.2 explains, the social consequence was large-scale land grabs and reservoir clearances, with compensation for landowners and no regard for hinterland communities. This finally became a political issue with Liverpool's flooding of the Tryweryn valley and Capel Celyn village in the 1960s. For the Welsh, Tryweryn epitomises the post-colonial relationship between England and Wales, which is unsurprisingly not an issue for the English.

If the power relations around reservoir construction are the same from the 1880s onwards, the outcomes are interestingly different. The period from the 1880s to the 1970s saw reservoir construction on an increasing scale. Then, from the mid-1970s (a decade or more before privatisation), reservoir construction stopped, and the privatised water companies have built no new reservoirs. Victorian reservoirs, as in the lower Elan Valley, are picturesque, but, as Chapter 6.3 explains, the greater part of current English reservoir capacity was added through large-scale new construction in the 1945–75 period. According to the data covering UK reservoirs, more than half of current capacity was added in the 1950–75 period; by 1975, 90% of current capacity had been installed and most of the rest was under construction. In outcome terms, what has to be explained is why post-1945 reservoir construction came to a juddering halt after the privatisation of water.

Understanding why construction stopped and why it is now being resumed on a project-by-project, privately financed basis moves us closer to action, because it clarifies how power works and whose power dominates. To see this, we have to look beyond the revenue constraint, which was (and is) a real limit. The public water authorities were starved of Treasury funding in the 1980s, partly because recent construction had expanded

Introduction

reservoir capacity and because privatisation was pending. Then, as Chapter 2.2 explains, under revenue constraint in privatised water there was no money for any kind of large-scale enhancement expenditure, after some initial upgrading of sewage disposal in the 1990s. However, there is a more political explanation of how and why the privatised industry accepted revenue constraints, which Chapter 6.3 attributes to a shift in power which empowered the finance sector and financial elites.

As we argue, what happened in the 1980s was all-change, with a new alignment of elites and the elimination of external pressures from organised labour. Political, technical and financial elites get things done by working together in a power configuration which involves intra-elite relations of hierarchy, competition and cooperation. In water, the agenda and policy pathway is determined by the internal balance of elite forces within the power configuration. Water privatisation was initiated by Thatcherite elites in Westminster, and technocratic engineering elites were demoted, as they could no longer access funding for infrastructure expansion. Meanwhile, financial elites were installed at one remove as the dominant group. For thirty years, stock market, bond market and financial consortia made easy money from financially servicing companies that were building very little. The role of water company management was to restrict physical investment to patch-and-mend replacement. The priority was to ensure that companies remained 'investable'. Within all elite groups, company and sector investability was and is the generally accepted principle that empowers the finance sector. Thus, in a key 2024 speech, Steve Reed, as incoming Secretary of State at Defra, promised change with continuity: 'My immediate focus is to make sure, from now on, that customers and the environment always come first, and that the water sector can attract the investment that's needed.'[19]

Against this background, regulation appears as an irrelevance or a fig leaf. As Chapter 5.2 argues, Ofwat was inadequate to the challenges of the water system because it had a narrow brief related to balancing consumer prices and firm profitability. This 'economic regulator' was not equipped for the accounting task of policing financial engineers such as Macquarie, nor the engineering task of ensuring adequate enhancement investment. Instead Ofwat went along with low prices and investability. When Jonson Cox (Ofwat chair 2012–22) was looking back, he regarded one important measure of Ofwat's success as being that the industry 'continues to be attractive to private capital markets'.[20] As for the Environment Agency, Chapter 5.3 demonstrates that it had an impossibly broad brief and has ended up providing useful political cover for ministers dealing with the pressure of events. Better regulation and enforcement are, of course, necessary: regardless of governance or mission, all organisations are less likely to behave badly if responsibility for harm will cost them significant money or cause reputational damage. But regulatory reform is, like public ownership, a necessary corollary of other measures and not in itself sufficient to achieve progressive outcomes. Amid confused policy debate, regulatory reform can easily become a distraction which postpones going anywhere near the planning and coordination we now need.

The mess we are in and what to do

Focused and disruptive action is urgent because water is in a mess. The water companies are not dealing with the legacy

Introduction

of physical under-investment and the accumulation of debt. Indeed, under PR24 – the price and investment allowances determined by Ofwat to cover 2025–30 – the water companies that cannot repay more than £60 billion of existing debt are being allowed to issue new debt. In their draft PR24 business plans for these five years, the water companies proposed to raise £25.3 billion of new debt. As Chapter 6.2 explains, this is already an industry of over-capitalised zombie companies, which need restructuring with equity write-off and debt write-down, at which point, as part of a larger reform package, they could and should be taken into public ownership at relatively low cost. However, when investability maintains the privilege of the finance sector and its elites, what we will get is not company restructuring but projectification, with private finance for new large projects such as reservoirs, and distressed companies passed on to new owners for more financial engineering. New project companies are already being developed, and this is likely to feature prominently in the Cunliffe Commission's report as a good news story about getting things built, but large-scale projectification will be an unaccountable disaster. The single project focus – with projects nominated by water companies – undermines planning and coordination, and completed projects cannot be undone for anywhere from 25 to 125 years. While the government will pretend to be fixing problems, we will be in a worsening mess.

In pursuing foundational objectives such as a water system that delivers liveability and sustainability, there is always a primary choice between conciliating and challenging power, with a secondary choice about policy instruments. In our last book, *When Nothing Works* in 2023, we warned against the practically futile and environmentally irresponsible pursuit of economic growth and recommended starter, stealth and switch policies. These could be constructively and quickly pursued by a moderate social democratic government that accepted all kinds of framework constraints but was focused on improving liveability for citizens. The choice between conciliation and challenging of power depends on the analysis of specific circumstances and conjunctures. As we have noted, any analysis of the 'water system' such as the one in this book will inevitably be incomplete, because there are so many important aspects of water and we cannot do justice to all of them. But by following the threads through the chapters, we can show that in the present configuration – defined by the dominance of finance and the absence of a persistent and organised citizen opposition – little gets done unless it is to the advantage of finance. And what does get done in the next phase of projectification will be to all our disadvantage. Hence if we want to move towards the foundational water management set out in the previous section, the choice must be to challenge, not conciliate power, because little is achievable without disrupting the current power configuration. The question, then, is how should power be challenged?

The political left correctly diagnoses a problem with the deficit of democracy in water. However, introducing new representational and governance mechanisms – such as citizens' assemblies or democratic boards – will have limited effect without the development of a social movement of active citizens that can develop and drive a disruptive agenda. As Chapter 6.4 observes, regionally based water companies cannot reform household charging nor undertake catchment planning, and the intelligent recommendations of citizens' assemblies can be ignored by the political classes when finance remains powerful. If we are to

Introduction

escape these limits, Chapter 6.4 argues that we need a change driver in the form of a broad-based social movement that takes up the challenge of inventing new institutions as part of a larger campaign of disrupting the power configuration. The strategic aim should be to demote finance from master to servant. Even with an enlarged revenue base, those engaged in water management will need to draw down syndicated loans and issue bonds, but this kind of external finance should be used to serve an agenda and priorities that are defined outside the finance sector.

The question, then, is what kind of social movement would have the ambition and the heft to disrupt the power configuration? Here, Chapter 6.4 draws a distinction between a civil society movement that seeks improvement within the existing framework and a social movement that aims to change the framework. The successful European campaigns against water privatisation show that a social movement is possible. In England and Wales, a new water movement is developing around old and new, large and small NGOs, beyond formal party political organisations. The movement is heterogeneous and consists of groups with different agendas. Hence it is both civil and social. When water organisations come together on a lowest common denominator basis, their demands are those of a civil society movement for cleaner water through the enforcement of regulation and limiting profit. But at a more granular level, as Chapter 6.4 shows, there is serious radicalism about problems and solutions in old and new NGOs such as River Action and in partnerships such as Sustainable Solutions for Water and Nature (SSWAN). European water radicalism has a basis in the language of rights and commons, whereas English radicalism is more pragmatically based on a realist recognition of the gravity of the problem of water management. The question is whether many more organisations can rally around more radical common denominators and put together a new kind of politics that connects the long-term planning and coordination of the water system with democracy, recognises the need to reform water charging, and reinvents the institutions of water management for adaptation to climate change. A broad base of activists in focused organisations could build the leverage to disrupt the power configuration and add external pressure to take action to reform water management, not just to fix storm overflows. But that outcome depends on political education and organisation.

At the end of writing this book, we have learned something about the foundational economy as much as discovered what is beneath the murkiness of water. This book is a beginning in that it provides the first sustained, complexity-capturing account of a foundational system, and does so on the basis that the multiplicities of water cannot be captured through one set of concepts or represented in some kind of circuit diagram. Equally, this book resumes a stream of work on citizenship which figured in our 2018 *Foundational Economy* book and was then interrupted by the untimely death of Mick Moran. As we have noted, foundational economy is a developing way of thinking which includes normative, descriptive and analytic elements. In previous work we have identified two preconditions, and this book extends the analysis by identifying the third precondition of foundational freedom. The first condition is household liveability, which depends on residual income after paying for market essentials, access to essential services such as health, and access to social infrastructure. The second condition is sustainability, which includes collectively meeting the unavoidable costs of adapting to climate change, because even our best

Introduction

local efforts at responsible mitigation cannot compensate for global failure. The third condition relates to active citizenship, and this book underlines the importance of the progressive agency that comes from organised citizens claiming a role in deliberation and decision making in foundational matters such as water, and in disrupting rule of the few. When these three conditions are met, citizens will have the freedom to live lives they have reason to value.

Chapter 1

From sanitation to water management

Introduction

Chapter 1 frames the issues that are discussed and explored in the rest of this book: water is a not-so-simple business which is historically imbricated in popular and elite politics, and now faces the new challenge of water management under climate change.

Section 1.1 begins by dispelling the illusion that water is a simple business that should be easy to get right. In 1989 the water industry in England and Wales was privatised, and ten regional water and sewerage companies took over its assets and operations. The companies and their regulators can quite rightly be blamed for much that has since gone wrong. In fairness, they inherited a large, complex foundational system that was never going to be easy to manage and, more than three decades later, is now not easy to redirect. Collective provision of the household necessities of potable water supply and wastewater disposal

From sanitation to water management

requires connecting millions of households to an asset-heavy system. This is largely structured by the legacy effects of historical decisions and embedded in a complex natural water cycle that is now increasingly affected by climate change. Thus, we have inherited – but would not now choose – combined sewer systems that require large hydraulic treatment capacity because, as well as wastewater, they carry the rainwater surface runoff which is increasing due to wetter winters.

Section 1.2 presents a foundational history of water and sewerage since the 1850s, which is instructive if we are to understand today's problems and tomorrow's possibilities. Municipal provision had by the end of the nineteenth century solved the urban public health problem of faeces in drinking water, and there was continuing technical progress in the twentieth century. However, water and sewerage also have a much darker history of imbrication in popular and elite politics. In waste disposal, from the great stink of the 1850s to storm overflows in the 2020s, progress in reducing pollution is driven by episodes of popular indignation which typically encourage solutions that in turn create new problems. This was the case with Bazalgette's intercept London sewer system. In relation to water supply, from the 1880s to the 1960s political elites in the conurbations drove large-scale upland land grabs and reservoir clearances, with no regard for hinterland communities in the Lake District and Wales. The lesson of this history is that we must enlist active, informed citizens in doing the politics of water differently.

Murky water

> *Section 1.3* explains that the task of management and control is now complicated by the way that our water problems have changed. The old problem was sanitation, which involved tapping freely available water supplies through reservoirs and abstraction, and disposing of waste on the principle of doing less environmental harm, such as with the ending of sewage sludge dumping at sea in the 1990s. But the problem of sanitation is now increasingly folded into a new, larger problem of water management in a period of climate change. There is considerable year-by-year and season-by-season variation. But if we consider averages since the 1880s, the English temperature has already risen by 1.5 degrees, winters are 25% wetter and summers are 10% drier. The prospect of water shortages which will limit crop growing and house building in the south and east of England is real, while the Environment Agency calculates that heavy rainfall in intense bursts puts one in four properties at risk of flooding by 2050.

1.1 Not a simple business

This is a book about the water industry, because water is foundational – it involves collective provision of a basic necessity – in a double sense.[1] Water is absolutely necessary for individual life. A classic 1940s dehydration experiment suggested that humans cannot survive much more than three or four days without water.[2] Clean water and sanitation systems for waste disposal are also absolutely necessary for collective life in urban and rural societies. In the absence of sewerage systems, infectious diseases

From sanitation to water management

cull urban populations, while rural areas may be overlooked in the provision of essential services. The World Health Organization estimated in 2022 that at least 1.7 billion people globally use a drinking water source contaminated with faeces,[3] even though 'universal and equitable access to safe and affordable drinking water' by 2030 is the first target under the United Nations' Sustainable Development Goal 6.[4] If many do not have what we have taken for granted in the UK, there is no room for complacency. Between 1831 and 1865, 100,000 people died in four UK cholera epidemics caused by contaminated drinking water. Thus, the UN General Assembly has recognised that water and sanitation is not only a human right but is also essential to the realisation of all the other human rights.

The UN's language of human rights dates from the 1940s, but central and local state initiatives to organise water and sanitation systems go back a couple of millennia. Karl August Wittfogel exaggerated when he argued that in hydraulic societies, from ancient Egypt to pre-Columbian Mexico and Peru, state power was based on the management of water excess through flood control and of water shortage through irrigation.[5] But water and sanitation systems are important wherever the state recognises its responsibility to keep the population safe from civil harm, and the organisation of such systems typically increases the scope and power of the local or central state. This was the case in nineteenth-century Britain through local municipal initiatives that used cheap industrially produced materials such as cast iron pipes for water supply. Municipalities were important actors in the original provision of water and sewerage systems and, as we observe in Chapter 6.4, these services have been the object of social movements in some European countries aimed at reversing the privatisation of services in the late twentieth century.

Murky water

In England and Wales, there have been two major state initiatives since 1945 in the reorganisation of water and sanitation.

- In 1974, at the tail end of twenty-five years of post-war planning, a Conservative government enacted, and its Labour successor implemented, the rationalisation of fragmented, mainly municipal, provision in England and Wales. Ten regional water authorities were created, each with near-monopoly responsibility for water supply and sanitation in a river catchment area.[6] These authorities had a short institutional life of some fifteen years, but they remain important because they created the operating framework and footprint in which the industry continues to operate. This combines an aspiration for integrated river basin management with a disintegrated national system of management, with less than 5% of water traded inter-regionally.
- In 1989, towards the end of Thatcher's sell-off of state corporate assets, the water authorities of England and Wales were privatised as ten standalone for-profit companies, which were initially quoted on the stock market. There was an element of chutzpah in this scheme for privatising a foundational necessity because, as *The Economist* observed, 'if you can privatise water, you can privatise anything'.[7] Many other countries have licensed private operators of publicly owned water and sanitation systems. But England and Wales were the first – and so far only – countries to transfer the assets *and* operations of a national water industry to private owners. The privatisers introduced the safeguard of economic regulation by creating a regulatory agency, the Water Services Regulation Authority (Ofwat). This regulator's brief was to ensure adequate returns for

investors while protecting consumers from price gouging by monopoly private suppliers that would not necessarily act in the public interest.

This new settlement was launched in 1989 in a slim White Paper,[8] which promised benefits for all stakeholders – and for the environment – by asserting that for-profit management drawing on private capital markets would be better in every way. For example, customers were offered 'the prospect of higher standards, greater efficiency in the provision of services, a charging policy designed to pass on efficiency savings and keep bills down and the opportunity to hold shares in the undertaking'.[9] From the late 2010s it was increasingly clear that none of these benefits had been delivered, and everybody on the political left and right, including Conservative ministers from Michael Gove onwards, accepted that the privatised system had failed. They did so without engaging with the underlying business model problem analysed in this book: under the current and regressive system of charging customers, the industry does not have enough revenue to cover both the claims of capital (debt and equity) for financial returns and the claims of physical investment for system renewal.

Why were the underlying problems of the business model not engaged with? The dominant mainstream economics thinking not only created Ofwat as regulator but also fixed on the absence of market risks as proof that the industry had the simple tasks of water delivery and sewage disposal, which should be easy to get right. Indeed, the water industry is not fast-moving in terms of changing market demands, aggressive competition or changes in technology. It is a slow-moving, asset-intensive business where monopoly providers in a sheltered sector meet a stable locked-in

Murky water

demand at price levels set five years in advance by regulators. On this basis, the BBC's economics editor, Faisal Islam, concludes that water 'should be a simple clean business'.[10] The economist Dieter Helm, a leading academic commentator on utilities, notes:

> The provision of water and the collection and disposal of sewage are amongst the simplest of tasks to define. The former needs a supply (rain, either collected in the past in aquifers or current rainfall), water treatment works, to make sure the water is fit to drink, and pipes. Because water has to be instantaneously and continuously supplied, some storage is needed. When it comes to sewage, it is the reverse: a system of pipes to collect the stuff, sewage treatment works and natural capital capacity to deal with it, and then a disposal system.[11]

The tasks may be simple to define, but their execution is anything but straightforward. First, drinking water delivery and sewage disposal requires a huge physical collection and distribution system, which is technically complicated and powerfully overdetermined by the legacy effects of earlier decisions. The ten major water and wastewater companies have 293,000 km of water mains and pipes and 576,000 km of sewage pipes and mains.[12] The combined length of their mains and pipes is roughly four times greater than the total length of all the roads in England and Wales.[13] Much of the flow of water and sewage has to be pumped, with 3,869 water pumping stations to maintain water pressure and 31,756 sewage pumping stations raising sewage between levels.[14] By design, Milton Keynes has a separate sewerage system, but most of the rest of our old and new towns and cities have legacy, combined systems which carry wastewater from houses and rainwater from road drains.[15] Combined systems require generous hydraulic capacity in treatment works to cope with heavy rain. After years of under-investment the system does not

From sanitation to water management

have enough hydraulic capacity, so that spillages of untreated sewage from 15,000 storm overflow systems have become far from the exceptional occurrences that were originally intended.

Second, water and sewerage infrastructure, like all critical infrastructures which are essential for the functioning of society, requires a complex set of governance arrangements to help to reconcile the interests of several groups of stakeholders, including the water companies, their regulators and government departments, plus the households and businesses who pay the bills that fund their services, and the general public interest in a good-quality environment. As with all infrastructures, there are also inter-temporal issues related to anticipating challenges and needs and investing appropriately. There is always a risk that current decision makers will reflect short- rather than long-term priorities.

Consider, for example, water mains replacement rates. Around one-fifth of the clean water running through the pipes is lost to leakage,[16] yet as Chapter 3.1 shows, only 18% of the water mains network has been refurbished or renewed since 2001; at this rate it would take 150 years to refurbish the whole stock. The replacement rate was not determined by technical calculations about such things as the limited life expectancy of cast iron pipes installed before the 1980s; it was a default political decision jointly taken by the water companies and their regulator. For two decades after 2000, the companies and Ofwat shared the tacit objective of limiting replacement investment. This was because under the existing charging system more costly investment would increase household bills and provoke political opposition, especially as the burden falls regressively on households in the bottom half of the income distribution, as we explore in Chapter 4.2. The same decision makers are, as Chapter 3.1 shows, responsible for the replacement of only 6% of the sewerage mains network since 2001, and, at this rate, it will take over 350 years to refurbish the whole stock.

Replacement rates have until very recently not been publicly debated or disclosed. But things have changed in the mid-2020s, with historical under-investment (especially in water supply and sewerage treatment) increasingly recognised as a problem. That is above all because civil society organisation has created and empowered new activist pressure groups that have disrupted the balance of power. This was dramatised when Steve Reed, as incoming Secretary of State for Environment, Food and Rural Affairs, announced the Cunliffe Commission review into the water sector in 2024. The Defra press release carried Reed's statement and, beneath this, six encomia for the launch of the review by the chief executive officers of Afonydd Cymru, Wildlife and Countryside Link, Angling Trust, Rivers Trust and Waterwise, plus the policy director of the Wildlife Trusts.[17] These NGOs are part of a vocal policy community which barely existed or was less organised at the time of privatisation in 1989, and which now has expectations of improved river and coastal water quality. This community's disappointment with slow progress after the Cunliffe review will create new problems for the government. And disappointment is likely because, beneath the Secretary of State's statement and above the environmentalists' encomia, the first endorsement for the establishment of the Cunliffe review comes from the CEO of the Global Infrastructure Investment Association.

A third complication, which undermines the economists' view of water and sewerage as a simple business, is the strong reliance of this infrastructure on complex natural processes, notably the water cycle and the capacity of the natural environment to deal

From sanitation to water management

with waste.[18] In highly simplified form, the water cycle can be represented as a continuous circular movement of water from the sea through condensation into clouds in the atmosphere, which then falls as rain on the earth, before returning to the sea. In practice, the water cycle ties together all the elements in our planet's climate system: air, clouds, sea, lakes and rivers, vegetation, snowpack and glaciers.

The central point is that in the UK, a stable and benign water cycle could be taken for granted in generating water provision in the nineteenth century and for most of the twentieth century. Until recently, the local climate mostly provided abundant water, which only had to be captured, mainly by reservoirs in the north of England and abstraction in the south. As we will see in more detail later in section 3 of this chapter, as global policies for mitigation of emissions are failing, climate change is upsetting established water cycles across the world. In the UK, the new water cycle will bring more problems of water excess in wetter winters and water shortage in drier summers. These will require a deep adaptation of the water and sewerage infrastructure and innovations in land and water management practices. This is not only to ensure continuation of services – clean water provision and wastewater treatment – but to reflect the interconnectedness of water and sewerage management as a corporate business with much wider processes and relations.

Beside adaptation to a new water cycle, the water and sewerage infrastructure will require improvements to shift from being polluting to being regenerative. As will be explored in the next section of this chapter, for much of its modern history water and sewerage infrastructure was constructed on the assumption that the natural environment was an unproblematic source of water, as well as a sink for waste. Before we go into the details, we invite the reader to pause for a moment to consider this basic fact: sewerage treatment prior to discharge into watercourses was introduced in the UK from the 1880s, but the dumping at sea of UK sewage sludge residues from many major cities continued for another hundred years until the early 1990s, and was discontinued only under pressure from EU directives.

These complexities – the sheer size of the infrastructure, its complex governance, and its reliance on natural processes – show how water is not a simple business. Acknowledging these complexities also has implications for how we manage the infrastructure going forward. As we will explore in the book, responding to these challenges will require new investment and imply high collective costs. All this is happening in a phase of economic stagnation and a prolonged liveability crisis; this limits the possibility of raising the revenue required for upgrading the infrastructure when household finances are already squeezed.

1.2 Sanitation's technical and political history

From the nineteenth century the history of water and sewerage is conventionally framed as a story of progress from dark to light, with sanitation delivering clean drinking water and sewage disposal to solve an urban public health problem. In standard accounts, the focus is on municipal initiatives, heroic civil engineers, and the massive new structures that they left behind. Between 1858 and 1870 London gained a new sewerage network with the construction of 132 km of new intercept sewers, backed by the rebuilding of 266 km of old main sewers and 1,770 km of local sewers. Manchester after 1894 drew its main water supply from the Thirlmere reservoir, 150 km north in the Lake

From sanitation to water management

District, which can supply up to 227 million litres of clean water per day.[19] And in public health terms, the result was a problem solved: by the end of the nineteenth century the contamination of drinking water with faeces was a problem of the past. The point was proved by London's fourth cholera outbreak in 1866, which was confined to areas that had not yet been reached by the new intercept system.[20]

In deconstructing all of this and presenting a foundational history, we must distinguish between the technical and political history of water and sewerage. Looking backwards, the technical history is important because the application of available techniques shapes the socio-economic achievements of any one period and structures the possibilities of the next. But if we are interested in understanding today's problems and tomorrow's possibilities, the political history of improvement is foundationally more instructive regarding the drivers of change and its social consequences. In this section, we will consider in turn the technical history and then the political history of water and sewerage.

The technical history leads us to advances in construction and treatment. In dam construction, from Lake Vyrnwy in the 1880s to Capel Celyn in 1965, successive generations of civil engineers developed increasingly sophisticated and technically elegant methods of construction and infill for gravity dams, where the weight of the dam resists the force of stored water. Haweswater, completed in 1940 and now supplying 2.5 million households in north-west England, was the world's first hollow buttress dam with an inclined, jointed retaining wall supported by 44 buttresses.[21] But equally important and less noticed are the innovations in treatment technologies. This started with James Simpson and slow sand filtration, which was being used

41

Murky water

on a large scale to purify London water by the mid-1850s, and continued with the development of high-flow, rapid gravity filtration from the 1890s.[22] As sewage treatment came to be accepted from the 1880s, so increasing attention focused on process improvements. The breakthrough was the discovery of the activated sludge process in 1913 by Arden and Lockett at Manchester's Davyhulme sewerage works.[23] This made obtaining a less toxic sludge technically and economically feasible. With continuous development up to the present day, activated sludge treatment became (and remains) the global industry standard.

The advances in construction and treatment continued into the 1970s, and this is most obvious if we consider the chronology of megadam construction. Liverpool constructed Lake Vyrnwy reservoir in the 1880s, and Manchester and Birmingham constructed Thirlmere and the lower Elan Valley reservoirs respectively in the 1890s. But a much greater amount of extra capacity was added between 1945 and 1980, when planners anticipated continuously increasing industrial demand. The prototypes for post-war expansion of supply were Haweswater (1929–40) for Manchester and Ladybower reservoir (1935–43) serving the cities of the East Midlands. The Claerwen reservoir (1946–52) more or less doubled Birmingham's water supply from the lower Elan Valley reservoirs, while the Clywedog dam (1963–67) regulated the flow of the Severn to allow abstraction for the West Midlands. The Llyn Celyn dam (1956–65) served Liverpool and the Wirral. This phase culminated with the construction of Kielder Water (1975–81), the largest artificial lake in the UK, which was designed to serve the industry of north-east England.

In the technical history of water and sewerage there is a glorious long century from the 1850s, when civil engineering thinking was dominant, and this is important because it typically led to an over-specification of materials and capacity in sewerage and water supply.

- The classic example of over-specification in sewerage is Bazalgette's intercept project. In terms of materials, Bazalgette pioneered the use of Portland cement, which was stronger and resisted water better than lime cement.[24] And on those sewers with a rapid gravity fall, he insisted on Staffordshire blues (not London stock bricks), because the more expensive harder brick would resist the scouring motion of the waste.[25] As for capacity, in estimating the necessary sewer pipe diameter, Bazalgette started by taking the estimated waste flow generated by the most densely populated district and then doubled the diameter. As he explained, 'We're only going to do this once and there's always the unforeseen.'[26] Thus, although in retrospect his technical choice of a combined sewerage system can be questioned, Bazalgette's system had the capacity to cope with a substantially larger twentieth-century London population.
- Over-specification of capacity in water supply is a constant in all the major dam constructions. With the Elan Valley project in the 1890s, Birmingham forecast increased household demand and then formally requested 'sufficient quantity to provide water for at least 50 years'.[27] Municipal planners in the 1960s all projected continual increases in the industrial demand for water over the coming decades, though these never materialised after the first Thatcher recession of 1981. This recession permanently reduced British manufacturing employment by roughly 20% and inaugurated a period of rapid deindustrialisation. Thus, the 71-billion-litre capacity

of Capel Celyn was not required by Liverpool, which sold the water on to other municipalities, while the 200-billion-litre capacity of Kielder Water has not so far been required.

Over-capacity and over-specification have had important unintended consequences in the period since privatisation. From around 1976, 'Treasury brain' and economics-based calculations inhibited new investment by the public water authorities. After the 1990s and a brief investment spurt to recognise the requirements of new EU directives, increasingly parsimonious financialised investors rationed investment. But the inherited excess reservoir capacity in the north (and the lack of effective controls on abstraction in the south) allowed privatised companies to tolerate leaks and postpone investment in new water supply for several decades. The companies were typically much closer to hydraulic capacity in sewage treatment, where improvement projects had always been less glamorous than large dam projects. The rollout of the activated sludge process was notably slow in interwar Britain. The Beckton treatment works in East London (which serves a population of four million) first adopted the process in 1932–35, with subsequent upgrading of capacity in 1959 and 1967.

If we are interested in the future not the past, the political history of water and sewerage is more foundationally instructive about the drivers of change and its social consequences. Water supply and sewerage systems are, of course, interconnected. With volume water supply from Thirlmere about to become available, in 1890 Manchester Corporation ruled that new housing should have flush toilets, which would increase water consumption. It went further in 1892 in ordering the retrofitting of flush toilets in existing rented houses, leading to a larger volume of waste

From sanitation to water management

to be conveyed away by a new intercept sewer system.[28] But for the purposes of foundational exposition, we can separate the histories of sewerage and clean water supply. In the story of sewerage, we highlight the driving role of periodic crises of public disgust about sewage disposal, which is generally out of sight and out of mind. In the history of water supply, we highlight the quasi-oligarchic role of elite decision making, with concomitant disregard for subaltern communities. These twin themes of crisis and elite decision making recur throughout this book, and we will return to them in the final chapter in thinking about how to renew the water system.

The current crisis with regard to storm overflows has several precursors and is only the latest of several episodes of public disgust that have driven improvements in sewage disposal. The 'great stink' in the hot summer of 1858 fed public revulsion about the use of the Thames as an open sewer for the central residential area of London. The discharge of sewage created foul smells which were believed to be threatening by Members of Parliament and others who accepted the miasma theory of disease. The episode is significant in two respects. First, public disgust created the crisis which gave Joseph Bazalgette, as chief engineer of the Metropolitan Board of Works, the mandate and resources to redesign the London sewerage system. Second, the episode shows that public disgust is often unfocused and can lead to doing the wrong thing, or the right thing for the wrong reason, as with miasma theory in 1858.

London's technical fix was an intercept system which did not solve the problem of environmental pollution but shifted the point of discharge and nuisance downriver, below the main population centre. The logic was the same in Manchester, which independently adopted 56 km of intercepts in the 1890s.[29]

Murky water

Bazalgette's newly constructed intercept sewers took surface water and waste from London's existing system using gravity assisted by pumping. They terminated in two large, covered storage reservoirs, at Beckton north of the Thames and Crossness south of the Thames,[30] with untreated sewage then being discharged from the reservoirs into the river on the ebb tide.[31] As a loyal employee of the Metropolitan Board of Works, Bazalgette had a less than heroic role in the mid-1870s, when he denied that his intercept sewage outfalls were creating offensive deposits on the banks of the tidal Thames.[32] This is a useful reminder that corporate entities and their employees can both mobilise resources for improvement and resist necessary change.

Change comes only when a disaster exposes a problem and stimulates another episode of public disgust. In 1878 more than 600 lives were lost when the passenger paddle steamer *Princess Alice* sank after a collision just below the point of sewage discharge into the Thames some 10 miles from central London. Many of those who died were not drowned but poisoned, because 75 million gallons of untreated sewage had been released one hour previously on the ebb tide.[33] Amid public outcry, a Royal Commission was appointed to advise on London's waste disposal as part of the political crisis management. The outcome after 1887 was the introduction of a precipitation process of sewage treatment which produced a toxic residue sludge that was disposed of by dumping it at sea. Again, the pollution problem was not solved but shifted further downstream, from the Beckton and Crossness outfalls to the approaches to the Thames estuary, some 12 to 15 miles off Foulness Point.

For the next century this offshoring of pollution was out of public sight and hence out of public mind, as more than one-quarter of all British sewage sludge was being dumped at sea.

From sanitation to water management

A succession of so-called 'Bovril boats' disposed of London's sewage sludge for more than one hundred years from 1887 to 1998, when a state-of-the-art sludge incinerator was built at Crossness.[34] The same practice was adopted by other major English conurbations, so that there was sludge dumping off the mouths of the Solent, Mersey, Clyde, Forth, Tyne and Humber.[35] As part of a larger campaign against dumping waste at sea, by the end of the 1980s Greenpeace protesters in inflatables were harassing the vessels then being used to dump London sludge.[36] But their campaign did not lead to widespread public disgust. Sludge dumping was finally outlawed by an EU directive because other North Sea facing countries had abolished the practice and saw no reason why the UK should continue. The implication is that broad-based, noisy civic action is necessary to shift established malpractices, a finding that resonates with the current protests about coastal and inland water quality by groups including Surfers Against Sewage and River Action.

The history of water supply is a different and more complicated story, or more exactly two stories. In the English midlands and north, the story is about mega reservoirs built at a substantial distance from the major conurbations of Birmingham, Manchester and Liverpool, using dams in the Lake District and mid-Wales valleys. In the south of England, the story is much more about abstraction. As we explain in the appendix to Chapter 5, most of London's water needs have been and are met by abstraction from the rivers Thames and Lea, with the balance made up by groundwater from aquifers. As space is limited, we will focus in this chapter on the midlands and northern story, which from the 1880s to the 1960s is about oligarchy, as municipal elites took decisions on behalf of their urban populations, with little regard for hinterland communities.

Murky water

From beginning to end, the story of reservoirs at a distance from conurbations is the same story of a self-confident municipal political class making decisions on behalf of their city populations. After obtaining an Act of Parliament, a major municipality could purchase land and subsequently flood a valley without any consultation or planning inquiry. To achieve the enabling Act, the interests of the major conurbations were pressed blithely and ruthlessly inside and outside the Commons and the Lords. With regard to Birmingham's scheme for flooding the Elan Valley, Joseph Chamberlain MP asserted in 1892 that it was all for the good because poor Welsh commoners and tenants 'will find much to their advantage in having a Corporation spending millions in their immediate neighbourhood'.[37] In relation to Manchester's scheme for taking water from Thirlmere in the Lake District, the Bishop of Manchester in 1879 firmly dismissed 'carping' objections on the grounds that 'two millions of people (in Manchester) had a right to the necessaries of life from any portion of England'.[38] Insofar as there was dissent about Thirlmere, it was an intra-elite matter: the Thirlmere Defence Association, founded in 1877 as the first 'politicised landscape preservation movement' in the UK, included Victorian luminaries such as Octavia Hill and John Ruskin.[39]

The social consequence was large-scale upland land grabs and reservoir clearances, with no regard for hinterland communities. Under the 1892 Water Act, Birmingham Corporation gained the right to acquire 45,000 acres in the Elan and Claerwen valleys and build dams which would displace nearly 300 local people.[40] Large (mainly English) owners of land and common rights were generously compensated; the largest Elan landowner got £150,000 (maybe £20 million in 2024 prices) for his upland sheep walks, including commons rights. But his Welsh tenants,

From sanitation to water management

commons beneficiaries and their service tradesmen got nothing. Some replacement housing was provided, as when Liverpool Corporation paternalistically rebuilt the drowned village of Llanwddyn downstream of the Lake Vyrnwy dam,[41] though such housing was usually occupied by incoming maintenance workers. Birmingham Corporation by 1911 employed just four locals from the Elan and Claerwen valleys in its local maintenance workforce of 90.[42]

The collision between urban needs and rural communities played out in a dramatic last act in the 1950s and 1960s, when the post-colonial relationship between England and Wales was dramatised by the drowning of a monoglot Welsh community to meet Liverpool's need for water through the construction of the Llyn Celyn dam. The community of Capel Celyn in the Tryweryn valley first learned that their village was to be drowned by reading about the plan in their local paper, a Welsh edition of the *Liverpool Daily Post*. They protested, but the outcome was mortifyingly ineffectual Welsh national outrage, with the relevant enabling Act passed in Westminster, even though 35 of 36 Welsh MPs voted against. Subsequently the slogan *Cofiwch Dryweryn* ('Remember Tryweryn') has come to symbolise not only the struggle against the decline of the Welsh language but also the struggle for a new Welsh national identity, because, in the words of the Manic Street Preachers song, 'we're not ready for drowning'.[43]

In English politics the issue of consequences for hinterland communities never surfaced because, from the mid-1970s onwards, the English stopped building reservoirs. But Tryweryn has a lesson which is important when we look ahead to the development of the water and sewerage system. One hundred years of increasing technical ingenuity in reservoir design was

Murky water

not matched by any development of political sensibility about groups excluded from the hierarchy of power, in communities seen as marginal. On this issue of power relations, we should not aim to go backwards to a supposed lost golden age but move forwards by doing things differently. We will return to this in the final chapter of the book when we consider the possibility of a new politics in water, where activists disrupt the prevailing power configuration and drive the renewal of the system *with* citizen involvement. This disruption is all the more necessary if we consider the implications of climate change, which we turn to in the next section.

1.3 Water management under climate change

The last section explored the sanitation problem definition – clean potable water and out-of-sight sewage disposal – which endured for more than a hundred years and could take the environment for granted. In particular, a reasonably benign water cycle could be assumed in a temperate country such as the UK, with a seemingly limitless source of water from rainfall, if captured in reservoirs and via abstraction, so that there was apparently no need for anything like a national water grid. At the same time, the environment more broadly could be treated as a sink for waste on the principle of doing gradually less harm, realised in slow steps with advances in sewage discharge and treatment. Meanwhile, as Bazalgette was building intercept sewers, the coal-burning, steam-powered Victorians started environmental processes of climate change and global warming. These were triggered above all by the carbon dioxide (CO_2) produced by burning all kinds of fossil fuels (coal, oil and gas) on a profligate

From sanitation to water management

scale throughout the twentieth century. In consequence, by the point of water privatisation in the 1980s, the environment was not an unproblematic background where improvement requires us only to do less harm in sewage disposal. Rather 'the environment' is becoming increasingly uncertain and unstable, requiring us to mitigate and adapt to climate change, which will bring increasingly acute challenges of managing both excess water and water shortage.

The problem of sanitation does not go away, as is apparent from current concerns about sewage discharges from storm overflows, but it is now increasingly folded into a new, larger problem of water management in the context of climate change. Most obviously, with wetter winters expected over the next few decades, the storm overflow problem will get worse unless we have natural solutions that slow the flow of water into rivers, as well as investment in hydraulic capacity at sewerage treatment works. In this section we explain why there has been a lag in recognising the problem of water management against a background of ineffectual action to mitigate climate change. After climate change had become the mainstream scientific consensus, the 2015 Paris Agreement proposed restraining temperature rises. A decade later, it is clear that mitigation in the UK has failed because of how it was globally conceived, as well as how it has been executed politically. Despite the year-on-year variability of the UK's temperate climate, the historical evidence of the past fifty years shows that climate change is already happening. Hence, as we will see in Chapter 3.3, while every effort should be made to develop more effective and globally coherent climate change mitigation, this is now not enough. Climate change adaptation is also increasingly and inescapably on the UK agenda, as elsewhere.

Murky water

The intellectual discovery of human-induced climate change and the emergence of a new scientific consensus is a classic story of how theoretical knowledge and new measurement techniques interact to slowly move marginal conjecture into the (global) mainstream of received scientific wisdom. In this case the received wisdom is about how emissions affect atmosphere and ground temperatures. As late as 1953, the eminent German meteorologist Fritz Moller dismissed the connection between greenhouse gases and rising temperatures as a 'very questionable' theory.[44] But the consensus shifted in the 1970s and 1980s with the development of computers and data processing that could test climate change theories and make projections of future trends based on past relations.[45] By the 1980s a new emerging interdisciplinary consensus on human-made climate change was reflected in the establishment of the Intergovernmental Panel on Climate Change (IPCC) in November 1988 under the auspices of the World Meteorological Organization and the UN, with a standing brief to assess the science related to climate change caused by human activities. These scientific breakthroughs in the understanding of human responsibility for climate change identified a collective global problem which required a political response.

The political response has come over twenty years in a series of meetings under the auspices of the IPCC and associated intergovernmental bodies. The Kyoto protocol of 1997 recognised climate change as a serious problem and set out nationally binding actions and targets which applied to a limited number of countries and were voted down in the US Senate. Amid difficult subsequent negotiations, the follow-on Copenhagen Accord was not legally binding and was not adopted universally. An IPCC report in early 2015 compellingly linked human actions to climate change, leading to the Paris Agreement of some 200 countries in the same year. This set out an overarching goal to hold 'the increase in the global average temperature to well below 2°C above pre-industrial levels' and to pursue efforts 'to limit the temperature increase to 1.5°C above pre-industrial levels' in the benchmark period of 1850–1900.[46]

The Paris Agreement objective of restricting global temperature increase to 1.5 degrees was based on scientific advice. The objective was to prevent breaching tipping points and moving into a period of large-scale, accelerating and nearly irreversible climate changes, such as those resulting from the melting of Arctic and Antarctic ice sheets and thawing of permafrost. The global upward curve of emissions has not so far been inflected downwards by mitigation measures and, in 2023, the UN Global Stocktake concluded that 'we are not on track to limit global warming to 1.5 degrees'.[47] The European Copernicus Service and the World Meteorological Organization both concluded that 2024 was the warmest year on record and the first year in which the global temperature rose more than 1.5 degrees above pre-industrial levels.[48] In retrospect, this failure was overdetermined globally by its intellectual framing as a transition project, and at UK level by inept political execution through net zero targets.

Globally, the project of emissions mitigation has been framed through a concept of energy transition which, as Jean-Baptiste Fressoz argues, rests on a naïve stages theory of energy dependence, whereby a world which has supposedly moved from coal to oil and gas will now transition to green and renewable energy.[49] The reality is that in a world of economic growth, new technologies such as wind and solar power are additional sources of energy, not replacements at the global level, and there

Murky water

is a symbiosis between old and new technologies. Thus, between 1980 and 2010, coal usage increased faster than in any previous period, so that coal increased its share in the energy mix at the expense of oil.[50] The imbrication of old and new technologies defines the carbon intensity of global economic growth. Growth and emissions are relatively decoupled because it is now possible to get increments of gross domestic product (GDP) with smaller increases in emissions. But absolute decoupling – growth without increasing emissions – which is what is needed to drive down emissions to sustainable levels, has not been achieved on a global level over the past thirty years, as Exhibit 1.1 indicates. Furthermore, research shows that given available technologies and government behaviours, absolute decoupling of emissions from economic growth is highly unlikely in the next thirty years, while

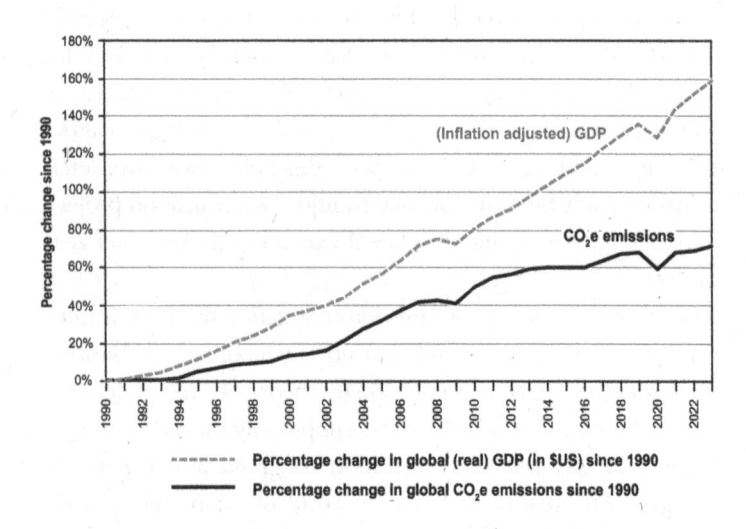

Exhibit 1.1 Change in global CO_2e emissions and world GDP (inflation-adjusted) since 1990[51]

From sanitation to water management

relative decoupling rates achieved by high-emissions countries fall far short of what is required to meet the Paris Agreement emission reduction objectives.[52]

At UK national level, the problems of muddled thinking were compounded by inept political execution. The Climate Change Act was originally passed in 2008 with near unanimous cross-party support, and has been widely represented as 'the world's first national framework legislation'.[53] As amended in 2019 by the May government, the Act set an economy-wide target of net zero greenhouse gas emissions by 2050.[54] But targeting net zero (not emissions) opened up lots of wriggle room around dodgy offset schemes, including tree planting,[55] and licensed optimism (or wishful thinking) about the low-cost scalability of technical fixes such as carbon capture and synthetic aviation fuels. More fundamentally, from 2008 onwards the Act allowed politicians to promise responsibility while postponing action on delivery. In the mid-2020s it is clear that the easy bit of UK emissions reduction has been done by phasing out coal-powered electricity generation. Further reductions in sectors such as transport and food production are not happening so far, and action to reduce their emissions would involve costs and inconvenience for households and businesses. The UK political consensus has now fractured, with right-wing media and politicians insisting that the public should not pay. In the 2024 General Election, the climate change headline in the Reform Party manifesto was 'net zero is crippling our economy', and the promise was that 'Reform UK will scrap net zero to cut bills and unleash growth'.[56]

The resulting debate on the need for and costs of mitigation distracted UK politicians and public from the two central issues in climate change. First, it allowed politicians to disagree about how to kickstart growth, while avoiding the issue that growth increases emissions. Second, it allowed experts to disagree about the avoidable costs of net zero,[57] while ignoring the issue that national adaptation to climate change is inescapable and will bring much larger and unavoidable costs.

If all this has largely not registered with the public, this is because the variability of the weather generally masks the effects of climate change. Britain does not have extreme weather events such as Hurricane Katrina in the United States in 2005, larger fires in the Amazon rainforest, and recurrent flooding which has displaced millions in Bangladesh. The one major post-war environmental disaster was the Great North Sea Flood of 1953, when a tidal surge led to the loss of 300 lives and the flooding of 160,000 acres along the English eastern coast.[58] Most other climate events are either rare and limited, such as the 'standpipe drought' of 1975–76, or relatively small-scale (though locally disruptive) such as the flooding on the Somerset Levels in 2013–14. Furthermore, in temperate England and Wales, slow change is not dramatised by a process such as retreating glaciers but is masked by the considerable variability of temperature and rainfall on a year-by-year basis. This means that, so far, declarations of drought or flooding tend to be intermittent or feel remote for most of the population, rather than appearing as a gradually increasing and disruptive process requiring mitigation.

But if we consider long- and short-run trends, the UK temperature is increasing slowly but in an accelerating way, and this will bring changes in rainfall. The UK is an offshore European island caught in endogenous processes of climate change as increasing temperatures upset the water cycle. With global and local increases in temperature, warmer air holds more moisture from the sea and there is increased evaporation from warmer soil and inland water surfaces. We already know that evaporation from reservoirs is increasing by 5.4% each decade.[59] The general pattern around the world is that land masses next to heating oceans have heavier rainfall and more extreme climatic events, manifest most obviously in more violent Atlantic hurricanes such as those that struck Florida in 2024. In the UK, the results are more complex and subtle.

Climate trends are complex because of significant annual variations and non-linear developments. We are not climate scientists who can make specialist assessments of the data, nor our own informed predictions about future developments. But we can look at the historical official data, and even with the crude approximation of a linear trend (Exhibits 1.2 and 1.3) we can establish some basic points that are relevant for thinking about the problem from the perspective of infrastructure policy.

- First, the long-run average temperature in England has already increased by some 1.5 degrees between 1880 and 2023, from 8.6 degrees to 10.2 degrees (Exhibit 1.2). While there is variation year-to-year, this variation was consistently below the trend line in the late nineteenth century and above the line since the 1980s. And the accumulating and accelerating increase has begun to deliver more palpable change in the last few decades. From 1884 to 1980 there were two years with average annual temperature above 10 degrees, but since 1980 there have been 22 such years.
- Second, if we consider the trends in precipitation (rain, snow, sleet and hail), we should note that these are more confusing because the range of year-to-year variation is much greater, and the pattern of variation appears more random. As Exhibit 1.3 shows, actual precipitation can often

Murky water

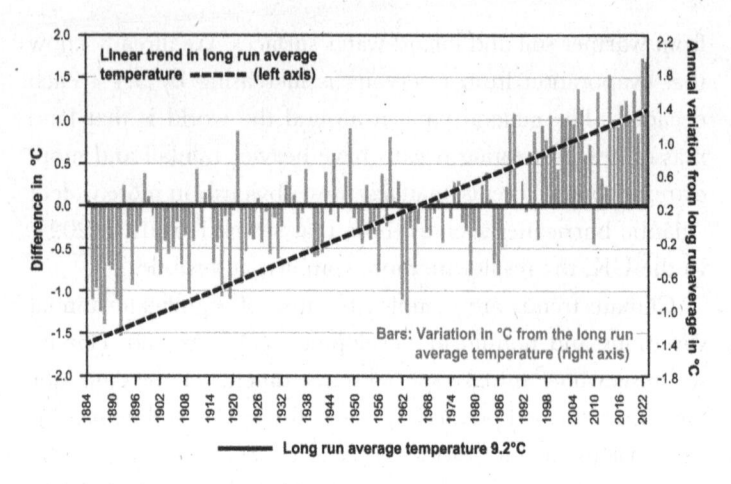

Exhibit 1.2 Annual variation in temperature relative to the long-run trend in England, 1884–2023[60]

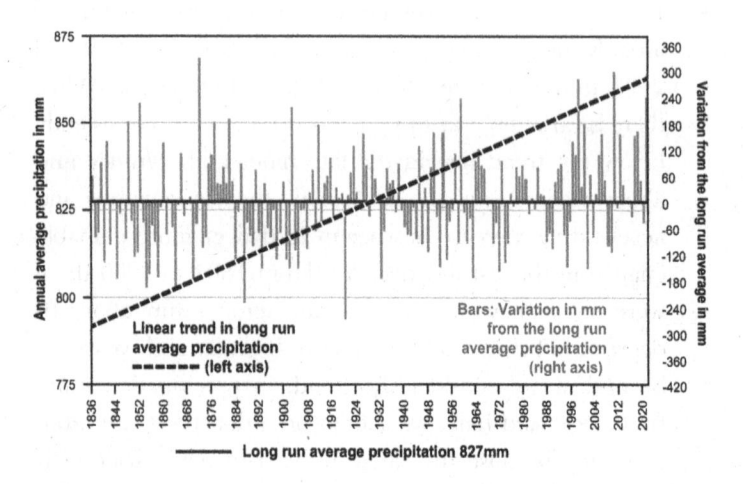

Exhibit 1.3 Annual variation in precipitation relative to the long-run trend in England, 1836–2023[61]

From sanitation to water management

be 30% higher or lower than the year before, as very wet years can follow very dry years and vice versa. Nonetheless, the long-term trend of annual precipitation shows a modest upward curve resulting from a higher frequency of exceptionally wet years and a lower frequency of exceptionally dry years in the recent period. If the 1836–1923 long-run average is the benchmark, in the period 1836 to 1980 precipitation was below this benchmark in six out of every ten years, while over the period 1980–2023 it was above this average in six out of every ten years.

National annual averages matter not least because, as we shall see in Chapter 3.3 where we consider UK Met Office projections, the upward curve of temperature increase can be extrapolated through climate models so that by the end of this century, under worst-case scenarios, temperatures right across the UK will have increased by at least 3 degrees. National averages in rainfall, however, conceal as much as they reveal. As Exhibit 1.3 shows, the gentle long-run increase in annual precipitation in England is the result of winters getting markedly wetter as summers get drier in the long term, and with autumn getting wetter since the 1950s. Exhibit 1.4 shows that winter precipitation has increased by some 25% from around 190 mm in the early 1880s to more than 240 mm in 2023; and summer precipitation has decreased by some 10% from 215 mm in the early 1880s to just under 200 mm in 2023.

The effects of wetter winters and drier summers are complicated by large regional differences in rainfall within the UK. Because rain clouds are typically carried on Atlantic westerly winds across England and the UK as a whole, the west and north get substantially more rainfall than the south and east.

Murky water

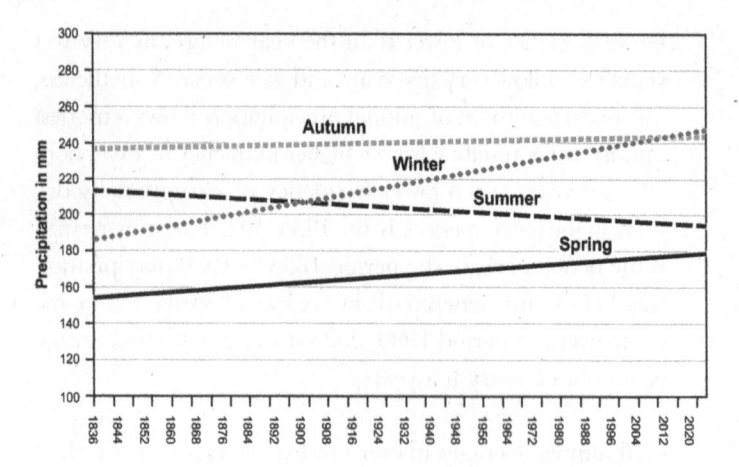

Exhibit 1.4 Seasonal linear trends in precipitation across England, 1836–2023[62]

Annual rainfall in the exposed western areas of England and Wales will continue to be twice what it is in East Anglia or south-east England. Considering England and Wales over the whole period from 1836 to 2023, north-west England and Wales are the wettest regions with a historical average of 1245 mm of annual rainfall. East Anglia is the driest region of England, with annual rainfall of 613 mm; south-east/central southern and east/north-east England follow behind with 760 and 768 mm respectively.[63] On this basis, wetter winters and drier summers bring a new water management problem.

Drier summers bring water shortages that will limit crop growing and house building in the south and east of England. There are already limits on house building in East Anglia because of a lack of water infrastructure.[64] More critically, as summers get drier there will be increased demand for irrigation for arable farming and horticulture, as well as greater scarcity of water.

From sanitation to water management

Existing abstraction rules which allow farmers and other commercial users to extract water from rivers and aquifers were designed in the 1960s for a period when water scarcity was not an issue. Without investment in more water storage and/or transfer, as well as reforms of abstraction rules,[65] there will be increased tensions about access to water, with significant food security and ecological consequences.

It is not simply the total rainfall that matters but also the intensity. As climate change warms the air, clouds hold more moisture drawn up from oceans and a greater number of larger rain droplets lead to more intense rainfall events.[66] Heavy rainfall in intense bursts presents greater flood risks and makes catchment management and urban modifications to slow the flow and prevent flooding much more important, especially in the north and west. There is already regular flooding of settlements such as Hebden Bridge[67] and Pontypridd[68] that are situated in narrow river valleys. And on the Environment Agency's 2024 assessment, some 6.3 million properties in England are currently at risk of flooding. Most of these properties are vulnerable not to river flooding but to surface water flooding caused by rainfall which cannot soak into the ground or be carried away by drains.[69] By 2050 the Environment Agency projects that one in four properties will be at risk of flooding, and the upcoming problems of surface water flooding are already leading insurance companies to red line several London districts.

It is clear that we have already entered into a new period where the disruptive changing environment needs to be integrated into our water and sewerage problem definition. The original focus on sanitation remains relevant and, as the outcry about the pollution of inland and coastal waters with raw sewage shows, this is not a problem that is easily solved, given rising

Murky water

public expectations. However, these concerns need to be set in a wider context. The scale and dimensions of our emerging water management problem going forward to the end of the twenty-first century present new challenges which are analysed in Chapter 3. Later in the book, we will consider the power configuration which keeps the failed privatised system going and effectively prevents our tackling issues of water management. But first we must look at the business model problems of the industry.

Chapter 2
The business model problem

Introduction

This chapter analyses the business model problem, which is that there is not enough revenue in the system to provide a return to finance capital and at the same time sustain the high level of physical investment that the asset-heavy water system requires.

Section 2.1 outlines the business model problem of the privatised industry. After covering operating costs, the cash surplus is not large enough to meet both 1) the external claims of those supplying finance capital in the form of equity and debt, and 2) the internal claims for capital expenditure in an asset-heavy activity with a voracious demand for physical investment. The intractable nature of the business model problem comes from the intersection of the two sets of claims. The activity characteristics of water and sewerage are just as important as the extractive irresponsibility of investors in driving the privatised industry towards financial and physical unsustainability, with a mountain of debt and chronic under-investment. Along with railways and the electricity grid, water and sewerage is one of three asset-heavy utilities which were, in business model terms, fundamentally unsuited to privatisation because the claims of finance capital and physical investment would exceed available revenue.

Section 2.2 relates the business model problem to a revenue constraint. Around three-quarters of industry revenue comes from domestic users. Water privatisation was politically sold with the promise of lower bills, while a regressive household charging system inhibited bill increases after an initial rise in the 1990s. The water companies and the regulator, Ofwat, became co-conspirators in keeping bills low in the 2000s and 2010s, regardless of the business model cost in increasing physical and financial unsustainability. Total water industry revenues were flat or decreasing from 2007 to 2023, when an increase in the number of connected households of around 185,000 per annum helped offset bills that were declining in inflation-adjusted terms. This was a backhanded gift, of course, since this also increased the number of households requiring water and sewerage services. After the problem of under-investment had been belatedly recognised, Ofwat and the water companies agreed that bills would have to go up in the PR24 price determination for 2025–30, and this increase represents only the first instalment of much higher charges which are inevitable given the legacy of under-investment. The *appendix* shows that the main policy focus on business users has been to spread competition and reduce their bills, with no high-level view of their role in funding the necessary investment.

Section 2.3 explores how ownership matters, and challenges the widespread assumption that extractive ownership causes the business model problem. Most of the companies that were privatised and listed on the stock market have since passed through various forms of private ownership by large utilities and investment consortia. Our comparisons show that ownership form does not make much difference to 1) financial extraction for all capital providers (equity and debt), or 2) productive commitment to physical investment. In the case of investment consortia, this is because the modus operandi of fund investors such as Macquarie is financial engineering through higher borrowing to produce a double-digit return for the owners, which – as in the case of Thames Water – wrecked the balance sheet. The most interesting case is the Welsh company Dŵr Cymru, which is a not-for-profit, bond-financed company (as the Labour Party proposed in its 2018 English water nationalisation plan). But Dŵr Cymru has been only marginally less extractive and productively committed than the for-profit companies, because bond finance is not a guarantee of low funding costs when it is tied to variable interest rates and price inflation. The virtue of not-for-profit bond finance is not that it solves the business model problem, but that it blocks opportunist financial engineering more effectively than any regulator.

2.1 The business model problem

Why has so much gone wrong financially and physically in the water industry? Most strikingly, an industry that was debt-free at privatisation has accumulated a huge debt mountain. By 2023 the water companies had collectively borrowed £68 billion, and Thames Water – the most financially distressed – was by early 2025 struggling under the burden of a reported £19 billion of debt and stumbling towards restructuring and debt write-down.[1] At the same time, the public is rightly indignant about incontinent storm overflows discharging untreated sewage into river courses and on to beaches. The problem is that privatised companies have not invested in infrastructure to increase sewerage treatment hydraulic capacity for the growing number of connected households. This is confirmed by the lead researcher of a rigorous academic study of storm overflows, who blames sewage overspills on 'the chronic under-capacity of the English wastewater systems'.[2] At least half of Thames Water sewerage treatment plants do not have the capacity to deal with incoming wastewater.[3]

The short answer to the question of why so much has gone wrong is that the water industry has a business model problem because there is not enough revenue in the industry to cover internal and external claims. The industry has a cash surplus after meeting its operating costs, including labour costs. But that cash surplus is not large enough to cover 1) external distributions to investors supplying financial capital in the form of equity and debt, and 2) the internal requirement for physical investment to replace existing capital equipment plus enhancement investment in increased capacity. The industry has covered this cash gap in two ways: first, by taking out more debt, which brings

The business model problem

in more cash every year but also increases the burden of interest payments in future years; second, by rationing physical investment with patch-and-mend strategies that dodge the requirement for enhancement investment. Thus, the long-term cost of the business model problem is a privatised industry that is financially and physically unsustainable.

The intractability of the water industry's business model under private ownership can be understood if we consider the relevant mechanics and magnitudes using a simple base and superstructure distinction. Every privately owned industry has a financial superstructure of external claims on its revenues from providers of equity and debt, whose expectations of return reflect current financial market realities and practices. It also has a base of internal claims on the same revenues which relate to the cost of running the day-to-day operations. These are specific to the activity and include replacement physical investment (leaving aside any questions of upgrade or capacity increase). These financial and operating claims are highly variable because every activity uses a different mix of capital, labour and bought-in materials to produce a product or service. The problem of the private water industry is, then, twofold. First, at superstructural level, external claims are required because financialised investors holding equity and debt in utilities will extract cash to obtain the going rate of return. Second, at base level, water is an extraordinarily asset-heavy activity, where the operating business has huge cash demands for replacement investment just to keep the existing infrastructure going. Both claims are very large, but the base claim for physical investment is actually larger than the superstructural claim for finance capital (even after the 1990s, when the water companies were not investing in infrastructure upgrades or capacity increase). Taking the revenue received from domestic and non-domestic customers between 1989 and 2023, 43 pence of each £1 of those revenues has been allocated to physical investment, and 35 pence has been claimed by finance capital in the form of shareholder dividends and interest payments on debt.

The problem with existing narratives of the water industry is that they focus on the financial superstructure and the irresponsible greed of investors. This irresponsibility is real enough, especially if we consider the conduct of Macquarie at Thames. But those preoccupied with irresponsibility have ignored the very distinctive activity specifics of the base, which is asset-heavy, with a voracious demand for replacement physical investment just to maintain existing infrastructure. An often told, powerful and irrefutable superstructure story of water relates to the irresponsible distribution of water profits with investment covered by debt. This account has, since the late 2010s, dominated mainstream media and political discussion on the centre-right as much as on the left, which originally highlighted financial extraction. But there is another, untold base story about the demands of physical investment and the limited scope for cost reduction in an asset-heavy, labour-light activity. The intractability of the business model problem comes from the intersection of these two sets of base and superstructure claims.

This section redresses the balance by giving activity base and finance superstructure equal status as drivers of water's business model problem. The issue of financial extraction and the often told superstructure story is covered in the first half of the section and highlights the expectations and behaviour of financialised investors. The very distinctive activity characteristics and the untold base story are covered in the second half of the section,

The business model problem

which highlights the replacement investment required to maintain a physical network of reservoirs, treatment works, pipes and mains. The two stories together give a clearer understanding of not only how and why the privatised water industry got into its current mess, but also what has to change on the revenue side if the industry is to fund the backlog of essential enhancement investment. The following section then explains the revenue constraint caused by the regulatory system which effectively capped price rises for two decades.

The superstructure story of financial extraction

The British public has always been resolutely opposed to the privatisation of water, which was a political elite project that directly and indirectly benefited financial elites. Three years before they was eventually privatised in 1986, unpopular proposals by the Thatcher government in the UK to turn the public water and sewerage authorities into privately owned companies were pulled from the government's legislative programme for fear they might damage its prospects in the upcoming general election.[4] In the period since 1989, opinion polls have consistently shown that a majority of the public favours re-nationalisation. YouGov runs an annual survey which tracks changing opinion on support for nationalising utilities. The 2024 poll showed that hostility to privatised water has increased in recent years: 82% of respondents in 2024 believed that 'water should be run in the public sector', as against 59% in 2017.[5] Against this background of sustained and increasing public hostility to the privatised water companies, it is not surprising that the industry has attracted the attention of financially literate researchers who have developed a story about financialised superstructure.

Murky water

The pioneer was Jean Shaoul with a prescient 1997 article on the early years of privatisation which highlighted the gap between promise and delivery, as well as identifying the many emerging problems.[6] Sustained critical attention came from Kate Bayliss of SOAS, University of London, working independently and with David Hall and Emanuele Lobina of Greenwich University. Hall for many years headed the Public Services International Research Unit at Greenwich, which has produced a continuous stream of critical studies of privatisation in the UK and elsewhere. Hall and Lobina's collaboration began with a 2001 briefing on UK water privatisation[7] and continued with their 2024 report *Clean Water: A Case for Public Ownership*.[8] After the 2008 financial crisis, Bayliss engaged with the financialisation literature and in 2014 produced a long working paper which set the water industry in this new frame,[9] and brought new researchers into the field. Around the same time, John Allen and Michael Pryke published a pioneering company-level analysis of financialisation at Thames Water, highlighting Macquarie's complex multi-level ownership structures and the way that debt-based financing of the company produced higher returns for shareholders.[10]

The problem for these critics in the mid-2010s was that their arguments were not getting traction. There was always a competing commentary by mainstream economists, some of whom had devised the regulatory framework and were more puzzled by, than critical of, water industry developments. A key figure here was the Oxford economist Dieter Helm. As early as 2003, Helm had noted with some puzzlement the trend towards much higher debt, 'but without the corresponding investment'. This had not been anticipated in the design of the post-privatisation

The business model problem

regulatory process, where it was intended that balance sheets would be used to finance physical investment, not financial returns to equity owners.[11] The breakthrough for the critics came in 2017 with a working paper authored by Bayliss and Hall which provided a succinct overview of the financial extraction story which had been developed at length in Bayliss's earlier 2014 case study.[12] The 2017 working paper presented an accessible accounting story about an irresponsible industry. Nine English water companies had distributed more than 95% of their post-tax profits as dividends between 2007 and 2016;[13] excessive distribution was accompanied by the accumulation of debt because the water companies were 'borrowing ever increasing amounts in order to pay for actual physical investment'.[14] This analysis was cited widely in the media, accepted by the economists, and finally picked up by politicians, so that it became the dominant mainstream account of the industry.

This story of irresponsible financialisation was intelligible for non-accountants and backed by relevant and irrefutable numbers. As Exhibit 2.1 shows, over the whole post-privatisation period from 1989 to 2023, the water companies have cumulatively distributed £82 billion as dividends in real 2023 prices.[15] This cumulative dividend is substantially larger than their cumulative £62 billion of post-tax profit, and the excess distribution has been funded by running down reserves by some £20 billion. Equally, an industry that was debt-free at privatisation has accumulated a significant debt mountain, reaching £68 billion by 2023.

This ownership story of irresponsible extraction by the water companies crossed over from radical reports into the political mainstream as a 'hiss the villain' narrative about greedy water

Murky water

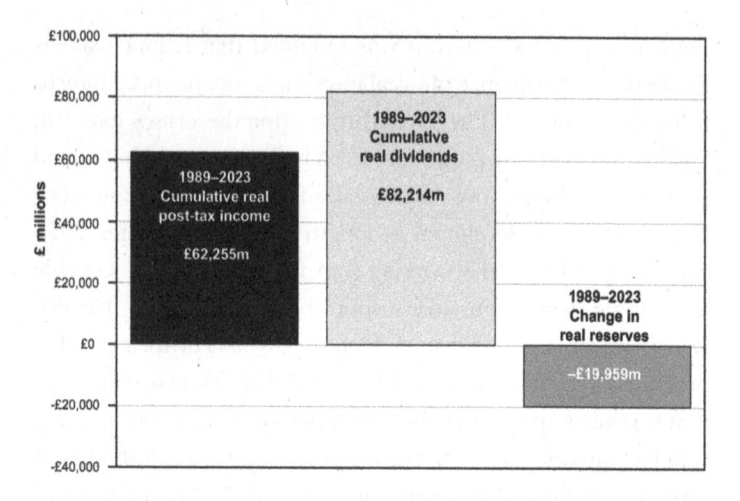

Exhibit 2.1 Post-tax income, dividends and change in reserves of water companies in England and Wales from 1989 to 2023, adjusted for inflation and presented in 2023 prices[16]

Bar chart showing:
- 1989–2023 Cumulative real post-tax income: £62,255m
- 1989–2023 Cumulative real dividends: £82,214m
- 1989–2023 Change in real reserves: −£19,959m

industry investors. In 2018 Michael Gove, as the senior Conservative minister responsible for the water industry, gave a hard-hitting speech to the City Conference of Water UK (the industry's trade association), in which he endorsed the criticism of financialisation.[17] The key media source then became the specialist reporting of Gill Plimmer in the *Financial Times*. Along with colleagues, she has over many years produced an informed event-by-event commentary on the financial troubles of the water industry, and has latterly connected this extraction story to the scandal of storm overflows with the argument that 'sewage spills highlight decades of underinvestment'.[18] It is significant that over many years, negative media stories about the water industry have figured more prominently in the *Financial Times* than in *The Guardian*.

The business model problem

Bayliss and Hall had a brilliant but limited success. The mainstream accepted their agreed facts but did not buy into the re-nationalisation argument. This was clear from Gove's key speech in 2018, when he asserted there was 'growing public concern' about the operation of the water industry. Gove cited Bayliss and Hall's data on dividend distributions as 95% of profits (without acknowledging the source) and added criticisms of tax avoidance and opaque financial structures. His radical conclusion was that 'some companies have been playing the system for the benefit of wealthy managers and owners, at the expense of consumers and the environment'.[19] But Gove maintained the longstanding Westminster Conservative position that privatisation was a *fait accompli*, and re-nationalisation would be 'a terrible backward step'. Instead, Gove argued for better regulation by Ofwat as it set prices and approved investment in the then upcoming price review, PR19, for the next five year period (2020–25).[20] Moreover, he asserted that regulation could and should rebalance priorities so that water companies would then be 'working as diligently on behalf of consumers and the natural world, as they are for their owners'.[21]

Against this background of agreed facts about irresponsible extraction – notwithstanding different views about appropriate forms of ownership – the activity characteristics of water were not brought into focus. This was partly because issues related to the different activity characteristics of various utilities were irrelevant and invisible for 2010s advocates of re-nationalisation, as they had been for the 1980s advocates of utility privatisation. Both were system-blind in promoting a preferred form of ownership as superior for all utilities (without considering activity differences between utilities).

Murky water

The base story about activity characteristics

The base story about the activity characteristics of water companies starts from the fact that industries use different combinations of assets, labour and other inputs to produce their products or services. A first key characteristic of the water and sewerage industry is that it requires a large stock of physical assets – land, buildings, plant and equipment – to operate. Regardless of whether the companies are state or privately owned and managed, the water industry is inescapably 'asset-heavy' or 'asset-intensive'.

This point emerges very clearly from Exhibit 2.2a, which relates 1) the stock of physical assets to 2) the sales revenue received from customers in a number of different industries, including other asset-heavy utilities (railways and electricity) and a sample of private firms engaged in different activities. In 2023, to generate £1 of sales revenues, English and Welsh water companies on average employed £6.50 of physical assets (including pipes, treatment works and reservoirs). On this measure, water is one of the most asset-heavy industries in the country. It sits between Network Rail (which operates the railway track, signals and many stations), where £8.62 of physical assets are required to generate £1 of sales revenues, and National Grid (which owns and operates the electricity transmission network in England and Wales), where 'just' £2.84 of assets are required to generate £1 of sales revenue. The average water and sewerage company is also around seven times more asset-intensive than BT (BT Group PLC, formerly British Telecom), whose cable networks reach every urban street in the UK, and which requires only 89 pence of physical assets to create each £1 of sales revenue. Major private firms which are not utilities are in a

The business model problem

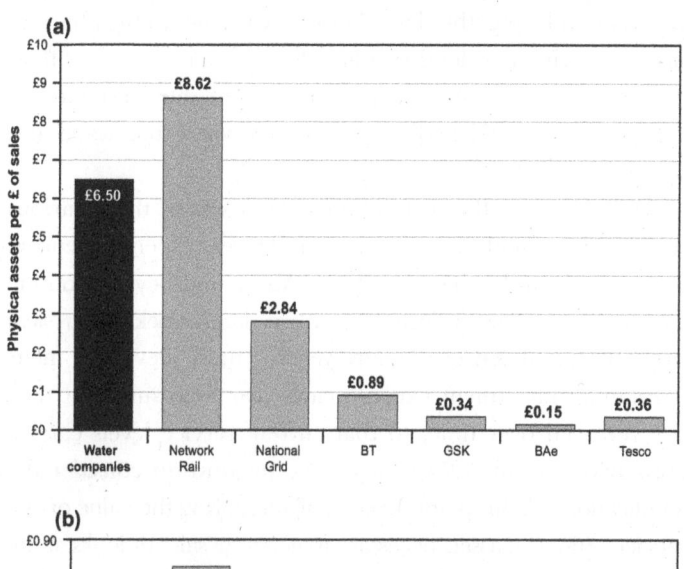

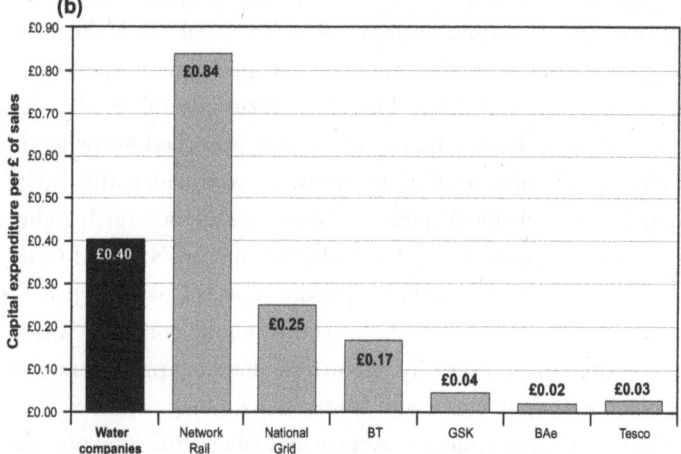

(a)

Physical assets per £ of sales:
- Water companies: £6.50
- Network Rail: £8.62
- National Grid: £2.84
- BT: £0.89
- GSK: £0.34
- BAe: £0.15
- Tesco: £0.36

(b)

Capital expenditure per £ of sales:
- Water companies: £0.40
- Network Rail: £0.84
- National Grid: £0.25
- BT: £0.17
- GSK: £0.04
- BAe: £0.02
- Tesco: £0.03

Exhibit 2.2 Analysis of asset intensity in various utilities and other businesses, 2012–23: (a) the value of physical assets used to generate each £1 of sales revenues; (b) the value of annual capital expenditure on physical assets per £1 of sales revenues[22]

Murky water

different and altogether less asset-heavy league. In big pharma, manufacturing and food retailing, firms usually have less than 40 pence of physical assets for every £1 of sales revenues, so that they are a striking 16 times less asset-heavy than water and sewerage.

Differences in the scale of physical assets in these various activities have implications for the year-by-year expenses incurred in replacing and enhancing them. Above and beyond routine maintenance, if an industry has a very large stock of physical assets accumulated over many years it must inevitably incur substantial investment expense each year to ensure that they are replaced over time, so that current service levels can be sustained. Exhibit 2.2b changes the measure to consider the implications of this point. Instead of measuring the value of the stock of physical assets necessary to generate sales, it looks at the average annual flow from sales revenue into capital expenditure between 2012 and 2023. This flow is burdensome in all three asset-heavy utilities: capital expenditure absorbed 40 pence in every £1 of sales revenue in the water companies; this is less than the remarkable 84 pence in Network Rail, but much higher than the 25 pence per £1 of sales revenue in National Grid. By way of contrast, capital expenditure at BT absorbs just 17 pence per £1 of sales revenue, even when the company was in the middle of an expensive network-enhancing programme of replacing copper wires with fibre. As for other activities from manufacturing to retail, the average annual investment in physical assets by leading firms in these sectors typically accounts for less than 5 pence of each £1 of sales revenues.

The huge spend on physical assets is all the more remarkable because in water – as in the two other asset-heavy utilities of

The business model problem

rail infrastructure and the electricity grid – investment spend on physical assets has been brutally rationed over several decades. Replacement investment has patched the existing system, while dodging costly large-scale enhancement investment that adds capacity and/or upgrades systems to provide *better* services. By 2024 the results were obvious in water and the other asset-heavy utilities. The privatised water industry has avoided mega-projects, most obviously by building no new reservoirs; and, as we show in Chapter 6.2, major enhancement projects going forward will not be funded by the water companies but under private finance arrangements using special purpose vehicles.[23] The other two asset-heavy utilities in Exhibit 2.2 have different problems related to legacy inadequacies with similar causes. Only 38% of Britain's mainline rail network is electrified[24] and some recent upgrades (as in South Wales) have electrified only part of the relevant main line. Meanwhile only four out of ten motorway service stations have fast chargers for battery electric vehicles, mainly because electricity grid supply connections are not available.[25] In all three cases, the limits of replacement patch and mend have been painfully exposed after a couple of decades.

The business model problems are compounded by a second characteristic of the water industry. This activity is not just asset-heavy, but also labour-light, so that the cost of other claims cannot be covered to some extent by reducing labour costs through headcount reduction and/or the degrading of wages and conditions. In water and sewerage, fluids move around the system largely without human assistance, and the role of a relatively small workforce is primarily to maintain structures and to provide customer service. Capital investment has generally not been directed towards taking a significant amount of labour out of the operating base activities. Critically, there was no change in technology that saved labour costs in water after 1989, as there was for example in telecoms with the move from the Strowger switch system to digital exchanges. In the ten years after the privatisation of BT in 1988, the workforce almost halved from 243,000 to 125,000, providing significant scope for cost reductions.[26]

These issues in water have been discussed in an abstract and inconclusive way by economists using their toolkit for measuring productivity increases. The water White Paper in 1989 was optimistic about efficiency gains, and economists have tried to measure total factor productivity, which is the gain in (quality-adjusted) output related to inputs of both labour and capital. An early study in 2001 by David Saal and David Parker found no improvement in the long-term growth rate of total factor productivity after privatisation.[27] A later Frontier Economics study commissioned by Water UK in 2017 described a privatised water trajectory of significant increases in productivity in the 1990s, deceleration in the 2000s, and stagnation in the 2010s.[28] These studies give a false impression of precision because their calculations depend on the unjustifiable assumptions that underpin the measures. More than half the cumulative early increase in total factor productivity in the 2017 Frontier Economics study was attributed to imputed increases in the quality of the water industry's output, which was a nonsensical assumption when at this point storm overflows were not being monitored.

If productivity is a problematic concept in relation to the water industry, it is more sensible to look within an accounting frame at the possibility of cost reduction by reducing employment. The scope for labour cost reduction was limited at the point of

The business model problem

privatisation because in 1989 employment costs accounted for only 19% of sales revenue in the operating business, compared with 37% for BT just after privatisation.[29] Since then, there has been no significant sustained reduction in the numbers employed: the industry in England and Wales employed 42,000 in 1989 and 42,999 in 2023. This reflects two developments. First, there had been significant headcount reductions before 1989, in the run-up to privatisation when the water authorities were under pressure to reduce costs so as to 'make the successful floatation more viable'.[30] Jean Shaoul calculates a 17% reduction in numbers employed from 1981 to 1989, partly because of the increased use of contractors.[31] Second, in more than thirty years since privatisation, the long-run trend is that industry headcount fell but then rose again. The initial fall was partly a result of reorganisation and changing industry boundaries, as when staff transferred from the old regional water authorities to the new environmental regulator, the National Rivers Authority (NRA). The subsequent rise after 2005[32] is likely to reflect greater requirements for staff to deal with regulatory affairs, including the ever more complicated price review process every five years, as well as the administration of the WaterSure and social tariff schemes discussed in Chapter 4.1.

The implication is that there has been no sustained scope for reducing labour costs as a way of offsetting the ongoing internal claims of capital expenditure for physical investment or the external finance capital claims for interest and dividends. In water (as in the other asset-heavy utilities of railways and the electricity grid), both sets of claims have to be met out of the cash surplus available after the operating expenses of paying the workforce and suppliers have been defrayed. The nature of the business model problem is clarified in Exhibit 2.3. This

Murky water

shows that in water, as in the two other asset-heavy utilities, the internal and external claims since privatisation (or restructuring in the case of rail infrastructure) add up to more than 100% of internally generated cashflow:[33] 137% in water, compared with 124% in the grid and 160% in Network Rail. This means that in all three of these utilities, the cashflow that originates from customer bills has to be topped up with new equity and/or debt.

The trajectory of all the privatised asset-heavy utilities is to move towards a spoilt balance sheet and financial unsustainability, as accumulating external finance capital claims only compound the problem of the irreducible, large, internal capital expenditure claims on the cashflow. Network Rail and the water companies both have £60 billion or more of debt, and National Grid has £45 billion of debt plus £10.2 billion of new equity after rights issues in 2010 and 2024.[34] Meanwhile, the aim of the chief

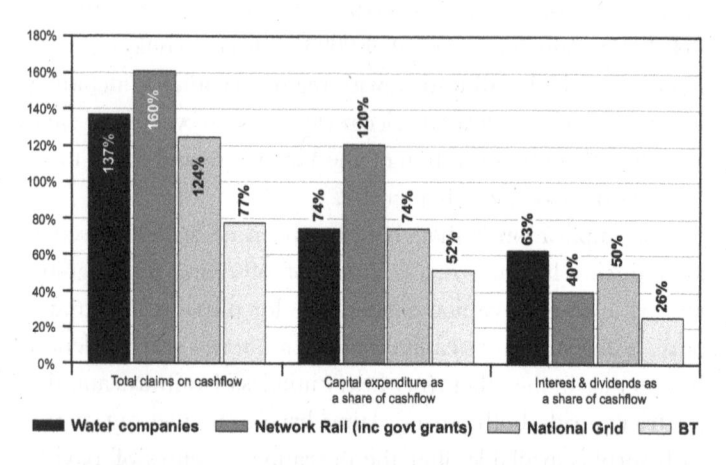

Exhibit 2.3 The distribution of operating cashflow for capital expenditures, interest and dividends among four utilities since their privatisation or creation[35]

The business model problem

finance officer of an asset-heavy utility company is to try to postpone the doom loop triggered by credit rating downgrades. This outcome will be quickly reached if the company borrows both to invest in assets and to meet the dividend and interest claims of shareholders and bondholders. The obvious tactic is to restrict physical investment so that it can be funded from internal cashflow, that is, the cash generated from the operating business. In the water companies – as in National Grid – total capital expenditure from 2012 to 2023 absorbed 74% of operating cashflow, while for Network Rail capital expenditure was at a high distress level of more than 100% of cashflow.

2.2 The revenue constraint

The immediate business model problem in water, as we have seen, is that the internal claims of physical investment and the external claims of finance together add up to more than the available cashflow. This in turn directly relates to a further business model issue, the sales revenue constraint. In an ordinary consumer market situation, private suppliers faced with a problem of this kind could increase the price of their goods or services to recover their costs, including whatever surplus is necessary to cover interest and dividend payments; or they might cut back their operations to remove higher cost components. Both of these options are complicated in water and sewerage. Water companies cannot retreat from some parts of their network and refuse to provide services in higher cost locations. Nor can they simply increase the price, because this is not set by the company but by the industry regulator in a framework that seeks to protect customers from the monopoly power of suppliers of essential water and sewerage services. The water regulator, Ofwat, has restricted bill increases so that, as we will show, total water industry revenues have been flat in real – inflation-adjusted – terms since 2007. Put simply, with bills held down in the 2000s and 2010s, no privatised water company has had a sustainable business model in which it could recover its costs by charging for its services.

The inhibitions in relation to raising household water bills have many causes, including as we shall see in Chapter 4 a regressive system of charging for water which limited bill increases because large increases would disproportionately burden low-income households. But historically, political expectations are also important because water privatisation was initially sold – and subsequently marketed – by politicians, regulators and water companies with the promise of low(er) household bills. This was the sweetener offered to a sceptical public when a clear majority were against water privatisation.[36] Thus, the first page of the water privatisation White Paper promised 'a system of economic regulation ... designed to ensure that the benefits of greater efficiency are systematically passed on to customers in the form of lower prices and better service than would otherwise have occurred'.[37] As we saw in the previous section, there was never any realistic prospect of efficiency increases. But that did not at the end of the 2010s prevent Ofwat and the water companies from opportunistically presenting bill reductions as an achievement which showed that privatisation was working (and at the same time glossing over the business model costs).

In 2019, thirty years after privatisation, Ofwat in its PR19 final determination promised lower household bills over the next five years while claiming that the industry could make the necessary physical investment. According to Ofwat, water

The business model problem

companies could 'invest for future generations and at the same time operate more efficiently and reduce bills', and so there would be a 12% real reduction in average household bills.[38] In the same year, looking back at the industry's achievements over the whole period since privatisation, the trade body Water UK admitted that household bills did not fall after 1989 and glossed over the fact that they actually rose in the 1990s when the privatised companies were investing to meet European Community directives on water quality. But Water UK could proudly claim that bills were low and had not risen in real terms since 2000, thanks to industry and regulator effort: 'at around £1 a day, average bills are broadly the same as 20 years ago once inflation is taken into account, and according to Ofwat they are around £120 less than they would have been without privatisation and tough independent regulation'.[39]

Fast forward just five years to 2024, and Ofwat (with ministerial support) and Water UK were taking very different positions, because they now recognised that bills had to go up significantly if the industry was to afford essential physical investment. In its PR24 final determination to cover investment and prices from 2025 to 2030, Ofwat insisted that the average household bill would have to go up by 36% in real terms, 'to help to finance essential investment in the water sector'.[40] Steve Reed, the Secretary of State, said that he supported the regulator's decision because higher bills 'will pay to fix crumbling infrastructure, which will dramatically reduce sewage spills and lead to cleaner rivers, lakes and seas'.[41] As for Water UK, it had shifted position so that now it wanted to blame Ofwat for holding bills down for twenty years. This emerged very clearly in a spat between Ofwat and Water UK after the regulator fined the water companies £158 million in 2024 for underperformance, and publicly said it wanted no more 'company excuses'.[42] The companies came back by blaming the regulator, with the CEO of Water UK arguing that 'Ofwat has prioritised keeping bills down and the result is chronic underinvestment, and we have to catch up'.[43]

Regulators, politicians and water companies have belatedly recognised that low bills have a business model cost in that they inhibit investment. The idea that Ofwat was directly responsible for keeping bills low and thus was at fault for restricting investment was by 2023 becoming the new received wisdom, as can be seen from the evidence of at least half a dozen expert witnesses to a House of Lords inquiry.[44] In its own evidence to that inquiry, Ofwat squirmed around the issue of how it had in the past balanced its two duties to protect consumer interests and ensure companies had adequate investment funding. The regulator asserted that it did not 'perceive a need to manage trade-offs in respect of [its] objectives'.[45] Its chief executive added rather lamely that he did not think 'that it is right to characterise Ofwat's focus as being on low bills'.[46] The Lords committee was unimpressed, insisted there were trade-offs, and reported 'the impression of some stakeholders that Ofwat has historically given more focus to a short-term desire to keep water bills low at the expense of long-term environmental and security of supply considerations'.[47]

This new consensus about lopsided trade-offs gets halfway there in defining the business model problems of the industry. It recognises that low bills, which directly benefited individual consumers, also created a revenue constraint at company and industry level which has limited investment. But it is not at all curious about what, other than expectations, determined the decades-long preference for low bills. As we demonstrate in Chapter 4.2, the existing household charging system is regressive,

The business model problem

so that low-income households pay more per person – both relatively and absolutely – than high-income households. The industry has had only patchy, small-scale schemes offering price rebates for some low-income consumers, and until recently has resisted a national social tariff. Under these conditions, higher bills would be seriously inequitable and meet political resistance. And so, Ofwat and the water companies became co-conspirators in maintaining low bills in successive price reviews for two decades after 2000.

Ofwat's determinations on household bills are decisive because in nine out of the ten major water companies, 75% or more of total revenue comes from residential customers.[48] The remainder is paid by commercial and public-sector users who, as noted in the appendix to this chapter, are not being asked to pay more or use less. United Utilities is the company with the least dependence on households, but even here domestic users contribute 71% of total revenues, and the domestic user contribution is at or above 80% or more at Anglian Water and Thames Water. It is difficult to calculate the average water bill per household in the early years after privatisation from public sources. However, the bill can be estimated[49] – using the number of dwellings as a proxy for the number of households – and the outcome presented in Exhibit 2.4 is striking. The customer would notice nominally higher bills because the average revenue per dwelling increases gently in most years, but the trend of inflation-adjusted (real) revenue per dwelling (based on 2023 prices) is quite different. Real revenue increased sharply in the first decade after privatisation when the industry was investing to meet European directives. Then, after a three-year dip in the early 2000s, real average revenue per dwelling recovered and lay between £600 and £650 for almost fifteen years. It then

Murky water

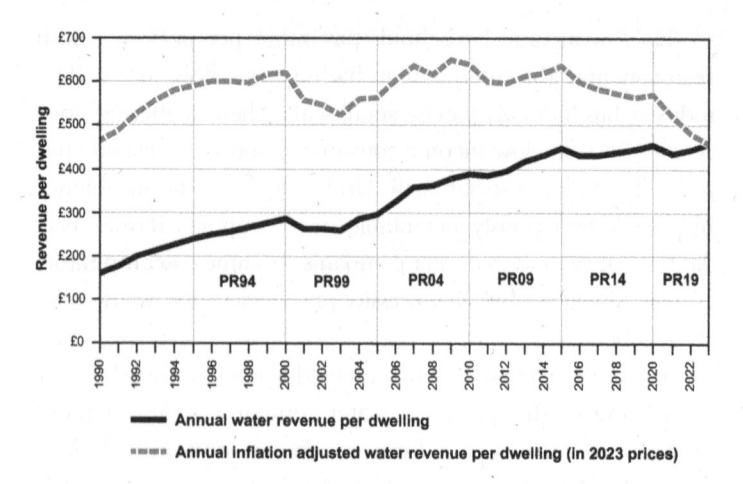

Exhibit 2.4 Average water company revenue per dwelling in England and Wales, 1990–2023[50]

dipped sharply again, falling to around £450 in real terms by 2023.

With real revenue per dwelling generally not increasing since 2000, the driver of the total national revenue increases shown in Exhibit 2.5 is an increase in the number of dwellings through new-build housing. In England and Wales, the number of dwellings has increased since water privatisation by 6.1 million, to reach 26.9 million by 2023. The new-build housing market is cyclical, but if we take the average across several cycles over more than thirty years, England and Wales typically add 185,000 new dwellings per annum. These additional dwellings contributed an estimated £2.8 billion per year of extra revenue by 2023 – based on the average bill – so that the water industry has benefited from 22.5% more revenue than it would have obtained without an increase in its customer base. While the industry

The business model problem

benefits financially from an increase in the number of customer charge points, this obviously means that there are more households requiring water and sewerage services. This extra revenue is thus a backhanded gift when the industry's investment has been focused on patching the system, not expanding capacity.

The increase in the number of connected households did not solve the problem of revenue constraint. Industry-wide in England and Wales, nominal total industry revenue has increased fairly steadily, but the trend is very different if we use the consumer price index (CPI)[51] to present the trend in real – inflation-adjusted – revenue since privatisation, as in Exhibit 2.5.

For the first decade after privatisation, real sales revenues nearly doubled, from £6 billion in 1989 to £11.5 billion by 2000 (in 2023 prices). This created headroom and made it much easier to manage the competing claims of physical investment

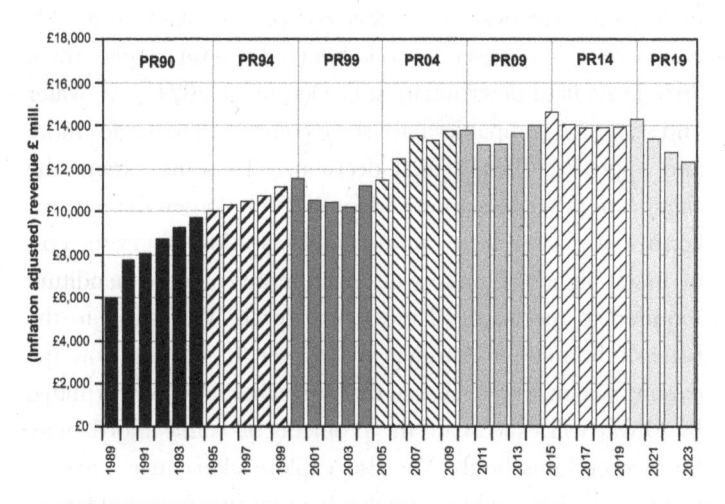

Exhibit 2.5 Total water company sales revenue (inflation-adjusted), categorised by price review period, 1989–2023[52]

and finance charges. During this period, the industry invested to meet European regulatory standards, which were cited by some commentators as an important part of the motivation for privatisation in 1989.[53] From the early 2000s this was followed by a distinct dip in real revenue, and then a thirteen-year plateau from 2007 to 2020, when real revenue was between £13.5 and £14.6 billion. Under the PR19 review, real revenue fell away sharply by nearly 15%, from £14.3 billion in 2020 to £12.3 billion in 2023. This represents a quite remarkable tightening of the revenue constraint in an industry which had by this stage piled up huge debt and had to meet increasing interest charges, while also facing accumulating problems caused by under-investment.

From this perspective, Ofwat's PR24 price determination process, which set investment and bills for 2025–30, was an important moment that signalled that the two-decade long period of low bills was over. The water companies asked for a 44% real increase in the average household bill and Ofwat gave them 36% in its final determination in December 2024. Five water and sewerage companies have since come back to ask for more and appealed Ofwat's final decision.[54] To some extent, the significant price rises approved for 2025–30 loosen the revenue constraint that we have outlined in this section, but at the cost of much higher bills and with much more capital expenditure required to meet agreed investment levels. We return to the issue of household charging in Chapter 4, showing how the existing system is regressive, so that the higher bills now permitted by Ofwat will exacerbate the problems of water affordability for poorer households. We also explore alternative ways of charging, which would ensure that high-income households paid more while low-income households paid less.

The business model problem

2.3 How ownership matters

So far, we have argued that there is an industry-wide business model problem in that the internal claims of physical investment and the external claims of finance capital cannot be met out of cashflow under the existing revenue constraint. This raises the question of whether and how ownership makes a difference. Are these business model conflicts palliated by more responsible forms of ownership, and can they be aggravated by more extractive ownership? Specifically, are stock-market-quoted public companies or multinational utilities more responsible owners than fund investors such as Macquarie? Does the removal of the profit motive with public or private not-for-profit ownership make a decisive difference? With just ten regionally based water and sewerage companies churning through different ownership forms it is not possible to have a rigorous experiment, but our comparisons later in this section show that the variations between different forms of ownership (including not-for-profit) in terms of extractive propensity or productive commitment are relatively small. This implies that there is no form of ownership (given the significant finance capital charges) which will resolve the business model conflicts. Put another way, rethinking water management will require more than a change of ownership.

The 1986 White Paper on water privatisation promised transformation, as water authorities were turned into public limited companies (PLCs) listed on the stock market. This would bring a new popular capitalism with 'the opportunity of wide ownership of shares both among employees and among local customers'.[55] Thus, all ten regional water authorities in England and Wales were initially privatised as PLCs, with shares listed on the London Stock Exchange. By 2025 only three London-listed water companies survived: United Utilities in north-west England, Severn Trent in the midlands, and Pennon, which owns South West Water. The vision of customers as shareholders was never realised and survives vestigially in just one company, South West Water, whose 'water share' scheme for customers has resulted in no more than one in sixteen of South West's customer households owning shares in the company.[56] The romantic vision of popular capitalism has been completely displaced by the realities of financialised capitalism.

The standalone publicly quoted water companies created in 1989 were established in two phases, as Exhibit 2.6 shows.

- In the first interim phase, from the mid-1990s to the early 2000s, there was a fashion for multi-utilities, which combined water and electricity activities, and included an incursion of multinational utilities into the UK. Hence North West Water merged with the electricity supply company Norweb in 1995, and Northumbrian Water was acquired by French utility company Lyonnaise in 1996. The result was disappointment as the multi-utilities folded, and only Wessex Water now survives under the multinational ownership of Malaysian YTL. But, in a classic capitalist story, the failure of these ownership changes to deliver on their promise created new opportunities.
- In a second settlement phase from 2002, public companies and conglomerate subsidiaries were taken private by infrastructure investment consortia and infrastructure investment funds. In 2002 First Aqua (a special purpose vehicle created by Citicorp) acquired Southern Water from Scottish Power; and in 2006 a Macquarie-led consortium, Kemble Water, acquired Thames Water from the German utility RWE.

The business model problem

Year	Company	Owner	Ownership type
1989	Anglian Water		PLC stock market listed
2006	Anglian Water	Osprey Consortium	Infrastructure Funds
1989	Dŵr Cyrmu		PLC stock market listed
1996	Dŵr Cyrmu	SWALEC	Multiutility
2000	Dŵr Cyrmu	Western Power Distribution	Multiutility
2000	Dŵr Cyrmu	Glas Cymru	Private limited by guarantee
1989	North West Water		PLC stock market listed
1995	North West Water	Merged with NORWEB to form United Utilities	Multiutility
2007	United Utilities	United Utilities divests electricity business	PLC stock market listed
1989	Northumbrian Water		PLC stock market listed
1995	Northumbrian Water	Lyonnaise des Eaux et de l'Eclairage (Lyonnaise)	Multiutility
2003	Northumbrian Water	75% sold to private investors	Private Investors
2011	Northumbrian Water	Cheung Kong Infrastructure Holdings	Infrastructure Funds
1989	Severn Trent		PLC stock market listed
1989	Southern Water		PLC stock market listed
1996	Southern Water	Scottish Power	Multiutility
2002	Southern Water	First Aqua	Private Investors
2007	Southern Water	Green Sands Investments Ltd	Private Investors
2021	Southern Water	Macquarie Asset Management (majority stake)	Infrastructure Fund
1989	South West Water		PLC stock market listed
1989	Thames Water		PLC stock market listed
2001	Thames Water	RWE	Multiutility
2006	Thames Water	Kemble Water Holdings Ltd (Macquarie)	Infrastructure Fund
2017	Thames Water	OMERS and the Kuwait Investment Authority	Pension/Sovereign Wealth Fund
1989	Wessex Water		PLC stock market listed
1998	Wessex Water	ENRON	Multiutility
2002	Wessex Water	YTL Power International of Malaysia	Multiutility
1989	Yorkshire Water		PLC stock market listed
2008	Yorkshire Water	Saltaire Water. Consortium of investment companies	Infrastructure Fund

Exhibit 2.6 Changes in ownership of water and sewerage companies since their privatisation[57]

Murky water

The outlier was Dŵr Cymru, a not-for-profit bond-financed water company which was created in 2001 out of the wreckage of Hyder, a Welsh multi-utility.

Against this background, it is not surprising that some centrist politicians, financial journalists and regulators now regret the rise of investor- or fund-owned companies and have argued for putting water companies back on the stock exchange as public limited companies. In a recent interview, George Eustice (the Conservative minister in charge of water from 2020 to 2022) lamented that, when publicly listed water companies were taken private by investment funds, they 'lost that connection with customers'; 'I'd like to see ... more of these water companies ... trying to get a listing once again'.[58] In a more hard-nosed way, the financial journalist Nils Pratley has repeatedly argued in *The Guardian* that 'the best long-term plan for Thames Water is to get it back on the stock market', because that would enforce transparency and accountability as well as limiting excessive leverage and opaque, tax-avoiding corporate structures.[59] In evidence to a House of Lords committee, Jonson Cox (Ofwat's chair from 2012 to 2022) regretted that 'there are not more publicly listed companies because it gives real visibility'.[60] The romantic political illusions about popular capitalism set out in the privatisation White Paper are more or less dead, but the preference for market discipline and visibility on the stock market lives on among some idealistic finance insiders.

The 64,000-dollar question, of course, is whether *public* ownership and management would make a difference to performance and outcomes. To answer this, we unfortunately cannot look to publicly owned Scottish Water as a test case. Since 2002 Scottish Water assets have been partly operated by English

The business model problem

privatised water companies under inherited private finance initiative (PFI) contracts; this was subsequently extended by Anglian Water's 2005 contract to service the non-business market.[61] In 2023 40–50% of Scottish wastewater was treated, and 80% of sludge was disposed of, under PFI contracts.[62] For now, Scottish Water is less a test case of public ownership and more an object lesson in the difficulty of undoing contractually embedded outsourcing. If the question is whether public ownership would make a difference, not-for-profit Dŵr Cymru is more relevant to the future ownership of the English companies than publicly owned Scottish Water. The Labour Party in 2018 proposed a nationalised company for English water which would be a bond-financed not-for-profit,[63] much like Glas Cymru, which owns and operates Dŵr Cymru's assets.

In these comparisons, we do not consider the possibility of a publicly financed water company which has 'free capital' in the form of state grants from taxpayer funds because we consider this exceedingly unlikely (and not clearly desirable) in the UK. Bayliss and Hall's 2017 argument for 'bringing water into public ownership' played with this possibility. They argued that stopping dividend distribution and interest payments would remove unnecessary costs and allow lower consumer bills, with households saving £100 per year or 25% of the 2017 bill.[64] From a business model point of view, and recognising the failure of enhancement investment, it would be more meaningful to say that stopping dividend and interest payments from 1989 to 2023 would have allowed a significant 80% increase in the size of the available investment fund. This would have been transformative in terms of the upgrading and replacement of infrastructure, but 'free capital' is no more than an interesting hypothetical because it is not going to happen.

Murky water

On any plausible scenario for public ownership of water in the UK, water companies would have to meet external finance capital claims for interest and/or dividends. This was certainly the case with the Labour Party's 2018 re-nationalisation proposal, which envisaged that the water industry's shareholder equity would be converted into debt.[65] Free capital is completely implausible in England and Wales in the 2020s when the UK Treasury has many urgent claims on taxpayer funds to maintain services such as health, education and defence. The Treasury would want any outstanding debt to sit on the water industry balance sheet and to be serviced by customers. If, by some miracle, Treasury grants provided 'free capital', this would remove finance claims and possibly relieve households of water charges, but at the cost of higher taxation and/or higher public debt. This would not necessarily remove either opacity or financial engineering, as charges are moved from one account to another, and debts are moved from one balance sheet to another. In Northern Ireland, households pay for their water as part of their domestic rates (without any itemised charge),[66] and any shift to taxation-based payments for water would encounter issues of regressivity in both local and national taxation.

With free capital cleared out of the way, we can turn to some simple financial comparisons of whether one form of ownership is more extractive and less committed to productive renewal than another. We propose two measures:

1. Dividends and interest paid as a share of sales revenues is a measure of the extent to which customer revenue is being skimmed to pay external providers of finance. A low percentage reading on this measure is an indicator of less extractive behaviour.

The business model problem

2   Capital expenditure as a share of operating cashflow is a measure of whether investment in physical assets from internal funds is being suppressed to facilitate external distribution. A high percentage reading on this measure is an indicator of more productive commitment.

Using these two measures, the best comparisons we can make are of two groups of companies and two individual companies. The long-run records of two groups of companies can be compared: the first is of the three stock-market-quoted companies that have existed since 1989, and the second is the five investor-owned companies, since they first passed into finance consortium ownership between 2002 and 2008. Two single company representatives of other forms of ownership can be analysed over the past two decades. Wessex has been under the multinational utility ownership of YTL since 2002, and Dŵr Cymru under Glas Cymru has been a bond-financed not-for-profit since the company's creation in 2001. These comparisons come with a health warning because the time periods used in the comparison do not correspond exactly, and individual companies serve regions of different sizes and character. But in the case of the two groups and the two companies we can at least make long-term comparisons of extractive behaviour and productive commitment.

The result in Exhibit 2.7 is that the group of five investor-owned companies and three stock-market-quoted PLCs look much the same. In terms of extraction, the investor-owned group distributed 36% of sales revenues as interest and dividends, while for the PLCs it was 35%. In terms of allocating operating cashflow to capital expenditure, the investor-owned group allocated 79% and the PLCs a similar 77%. The pattern is very clearly one of convergence between the two groups. Group averages, of course,

Murky water

(a)

Stock market listed PLC: 35.1%, 42.3%, 77.4%
Multiutility: 31.6%, 45.6%, 77.2%
Investor-owned: 36.0%, 44.9%, 80.9%
Not for profit: 29.6%, 38.3%, 67.9%

- Dividends and interest as a share of sales revenue
- Capital expenditure as a share of sales revenue
- Capital expenditure, dividends and interest as a share of sales revenue

(b)

Stock market listed PLC: 63.4%, 76.5%, 140.0%
Multiutility: 50.6%, 73.2%, 123.8%
Investor-owned: 63.1%, 78.7%, 141.9%
Not for profit: 64.1%, 82.9%, 147.0%

- Dividends and interest as a share of cashflow
- Capital expenditure as a share of cashflow
- Capital expenditure, dividends and interest as a share of cashflow

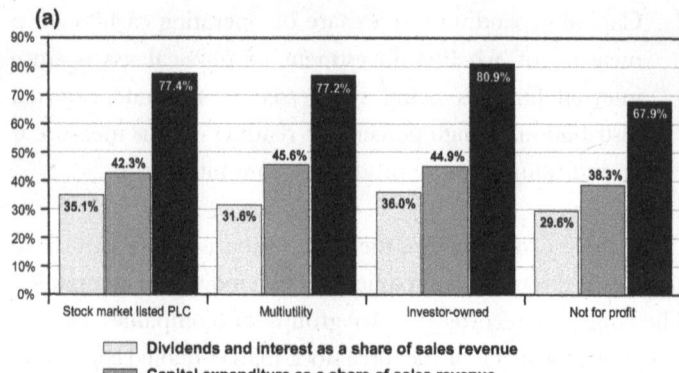

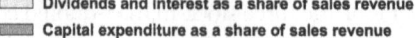

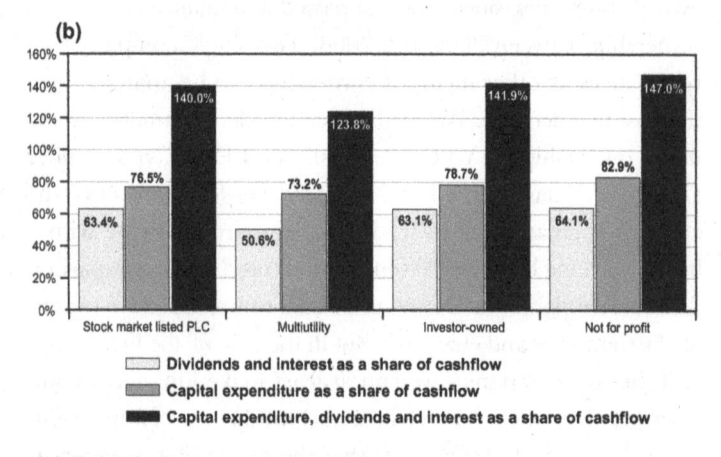

Exhibit 2.7 Claims made on sales revenues and operating cashflow in English and Welsh water companies, by ownership type: (a) capital expenditure, dividends and interest as a share of sales revenues; (b) capital expenditure, dividends and interest as a share of operating cashflow[67]

The business model problem

conceal considerable variation within groups. For example, using the distribution of sales revenue to dividends and interest, shown in Exhibit 2.7a, the three PLCs have distribution ratios ranging from 30% to 40%. And for this reason, we cannot place too much interpretative weight on the results of individual companies such as Wessex Water and Dŵr Cymru, which probably tell us as much about the peculiarities of these companies and the areas they serve as they do about the ownership form in any predictive sense.

It is notable, however, that Dŵr Cymru measures up as marginally better but not hugely different to the two multi-company groups on our two ratio tests. On the allocation of operating cashflow to capital expenditure (Exhibit 2.7b), not-for-profit Dŵr Cymru leads with 83%, but that is only 5% or so better than the two groups. In terms of distribution from turnover (Exhibit 2.7a), Dŵr Cymru distributes 30%, which is around 5% less than the PLCs and fund-owned groups.

A Macquarie-led consortium of equity investors bought Thames Water in 2006 for £8.5 billion, with £6.2 billion of the purchase price funded by issuing fixed interest bonds, so that equity investors put in just under £2 billion.[68] Over the course of this investment consortium's ownership, debt increased by a striking £9 billion to £11 billion, with no increase in equity from the investors. Increased leverage worked for Macquarie and the consortium's other equity investors. Over the consortium's decade of ownership, shareholders in the holding company Kemble obtained a gross 12–13% return on their part of £2.3 billion of equity.[69] The equity beneficiaries from 2006 to 2017 were, according to Macquarie, 'pension funds, sovereign wealth funds and insurance companies', which were mainly non-British and were investing the savings of the asset-rich classes in an increasingly wealth-based global capitalism. This financial engineering was simply a higher form of financial irresponsibility, because the cost of increasing leverage was a wrecked balance sheet and a growing burden of debt, which completely undermined the sustainability of Thames Water in the 2020s. But that was someone else's problem because Macquarie reduced its holding from 2011, and divested its remaining 25.3% ownership stake in 2017, leaving behind a debt-burdened company, as Exhibit 2.8 shows clearly. After Macquarie's exit, debt has continued to increase, and the next owners did not provide any new equity, which is now needed through restructuring in 2025.

As for Dŵr Cymru's small margin of superiority over the for-profit PLCs and investor-owned groups of firms, that should not be surprising. Not-for-profit bond finance, as realised by Glas Cymru in 2001 (and as proposed in Labour's 2018 nationalisation plan), does not guarantee low external financing costs because it brings into company financing all the happenstance of the outside world through exposure to highly variable interest rates and price inflation. Bond-financed companies can be lucky, most obviously when low interest rates prevail, but they cannot make their luck and can often be the victims of circumstance.

Glas Cymru, which owns Dŵr Cymru, is an unlucky bond-financed company which was initially financed when interest rates were high and more recently lost a bet on low inflation rates. Bond finance locks in market interest rates at the point of issue and again at the point of refinance because corporate bonds have a fixed lifespan. Prevailing interest rates, even for borrowers with good credit ratings, vary considerably between interlude periods of cheap money, as in the 1930s or 2010s. Glas Cymru was set up in 2001 with a £1.9 billion bond issue, which locked in interest rates much higher than those that

The business model problem

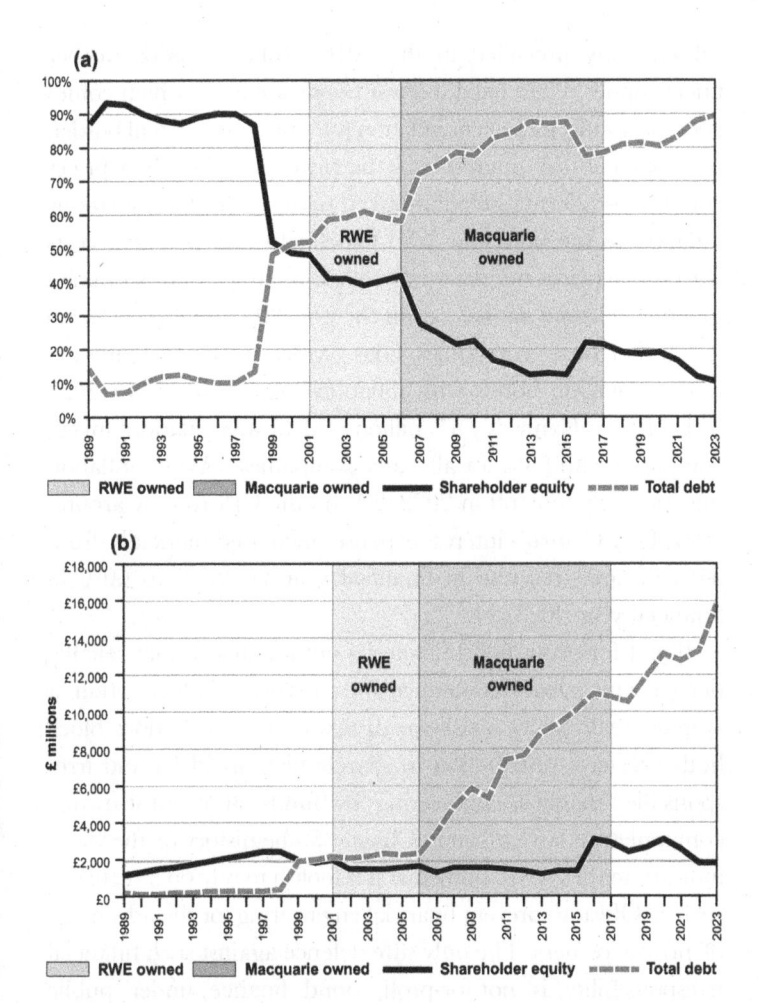

Exhibit 2.8 Thames Water equity and debt, 1989–2023: (a) equity and debt as a share of total capital; (b) value of shareholder equity and debt[70]

subsequently prevailed in the 2010s. Inflation adds further uncertainties. Plain fixed interest bonds are a bet which comes good on gentle, long-term inflation rates because the real burden of repayment of interest and the principal falls when prices rise. Bonds offering inflation-linked returns transfer the risk of inflation to the borrower, who benefits from lower debt costs with stable prices but pays more when price spikes or sustained inflation become an issue, as in the post-2022 period. More so than other water companies, Glas Cymru bet on low inflation rates by issuing bonds with inflation-linked returns. In 2023 85% of Glas Cymru's £4.5 billion of debt was inflation-linked, compared with 53% for all water companies.[71] When inflation unexpectedly took off in 2022–23 and the CPI rose by around 20%, Dŵr Cymru's interest expenses increased markedly, from 34% of total revenue in financial year 2022/23 to 60% in financial year 2023/24.[72]

If not-for-profit bond finance is not a panacea that reliably delivers cheap external finance and caps financial extraction, it is more realistically a risk worth taking because it does block both excessive distribution to shareholders in PLCs and irresponsible financial engineering by funds in investor-owned companies. As we explain in Chapter 5, the history of the water industry in the 2010s shows that it is foolish to rely on a regulator such as Ofwat to prevent financial engineering for the advantage of private owners. The only sure defence against such financial irresponsibility is not-for-profit bond finance under public ownership. If the bond finance bet on interest rates and price inflation then comes good for a while with lower financing costs, that will not solve the water industry's business model problem analysed in earlier sections of this chapter. This business model problem is that water and sewerage activities have a voracious

The business model problem

demand for capital expenditure from limited cashflow, under the revenue constraint established by the current system of household charging. Ever more debt and external borrowing is required regardless of ownership, and changing ownership is therefore a necessary but not sufficient condition of new water management, because in itself change of ownership is no more than a backstop control over financial engineering.

Appendix: business users and competition

Business customers of water companies are much less visible in public and policy discussion about the industry. This is partly a matter of history and path-dependence, since municipal water and sewerage companies were originally set up to serve domestic users, and this remains their primary function. This orientation is reinforced by long-chain global production and UK deindustrialisation which feed imports. Using water footprint calculations, the UK's 2011 consumption footprint from worldwide production is more than three times larger than the UK's production water footprint from UK domestic businesses.[73] Nevertheless, domestic business users are a small but significant part of the water company market, where they typically contribute no more than 20% of revenue to regional UK water companies. The extent to which large business users should pay more (relative to domestic users and small businesses) and use less water has not been the subject of policy discussion. In the absence of any policy shift that asks more of larger businesses, domestic consumers will most likely continue to bear the lion's share of the costs of renewing the industry and making good decades of under-investment.

Murky water

Wholesale prices are approved by Ofwat on an annual basis and are now increasing in line with domestic prices, so that there is an estimated average 42% increase in wholesale prices from 2025 to 2030.[74] The complication is that increases are being front-loaded by some cash-strapped water companies. Year-on-year, Southern is increasing its wholesale price for 2025–26 by 50%, and Thames is implementing a 32% price increase. Beyond this it is difficult to say which large users could and should pay more per unit because we have so little granular knowledge about who the big industrial users are. Regardless of unit price, all business users should be asked to use less and locate in ways that reflect water availability. Going forward, industrial use of water will exacerbate water shortage in the south-east of England, as agriculture increases irrigation demand and new water-intensive uses such as data centres are expanded.[75]

These issues have not been taken up because policymakers have been distracted by an attempt to promote competition and choice (of intermediary water resellers) for business users. As with many other attempts to encourage choice in privatised retail and wholesale markets, the results have been seriously disappointing and unsuccessful in their own terms. The persistence documented below shows mainly that the economists' dream of benefits from competition never dies.

Even supporters of water privatisation recognised that injecting competition into the market for water and sewerage services would be difficult. The White Paper that set out the case for privatisation in 1986 stopped some way short of proposing direct competition for customers.[76] Nonetheless, the idea that retail competition could be encouraged within a structure of regional monopolies that owned the infrastructure and were responsible

The business model problem

for upstream services[77] remained an aspiration for some and has been introduced in a limited way in several stages.

Competition for business users works through resellers providing customer services, billing, selling, metering and advice. Resellers buy water and/or sewerage services from incumbent regional providers at 'an agreed wholesale price and then on-sell to non-household customers'.[78] If all resellers pay the same regional wholesale price, lower prices and better service have to come from cost savings and innovation in the billing and customer service operations (which necessarily limits competition and price reduction). This competition for non-domestic users has been gradually extended in small steps from 2003. However, the take-up by business customers has been limited, and plans that households would also in due course be able to choose their supplier have been abandoned.

Initially, only businesses using more than 5 megalitres per year in England, or over 50 megalitres in Wales, could switch their water supplier. But from 2005 to 2013 only four business customers (and then only on some of their sites) changed their supplier. In Scotland, where retail competition was introduced for all non-domestic customers in 2008, around 5% had switched and around half had 'renegotiated with their existing supplier'.[79] The Labour government commissioned an 'independent review' of competition and innovation in the industry by Martin Cave, whose report was published in 2009.[80] This influenced the Water Act 2014 which allowed all non-domestic users to change their supplier of water and/or sewerage by removing the minimum size threshold, and encouraged more resellers to enter the market as providers of water and sewerage services.[81]

In the first full year of the expanded retail market, Ofwat reported positively on how it was working: around 10% of customers had switched in year one, with reported improvements in water efficiency and lower bills.[82] However, some years afterwards, in Ofwat's update for 2023–24, the benefit of the deregulated market seemed to be limited to a small number of high-use customers using switching to get lower prices.[83] Overall the results of retail competition have been disappointing. The National Audit Office (NAO) reported in 2020 that awareness, take-up and efficiency savings were all lower than expected, and few buyers were receiving water-efficiency advice or leakage-control services.[84]

The main benefit of such disappointment with competition in the business market is that this experience has discouraged experiments to introduce competition into the domestic market. The UK government originally planned that retail competition would eventually be extended to domestic customers. But Ofwat found in 2016 that this would be expensive to organise, customer take-up would most likely be limited, and projected reductions in customer bills would be small.[85] By 2019 any decision on rolling out competition for domestic customers[86] had been postponed, and this development is unlikely in the 2020s.

The Cunliffe Commission's Call for Evidence report in 2025 noted that most business customers have never switched supplier and that bulk tariffs provide no incentives to reduce water use. It did not consult on expanding competition to domestic consumers but did seek views on how existing competition mechanisms could work better.[87] The dream of competition lives on as a minor distraction from the issues that matter.

Chapter 3

Much more investment required

Introduction

This chapter shows how the water and sewerage system needs much more investment after decades of patch and mend by the privatised companies. While everybody now acknowledges the historical under-investment, the industry is promising more than it will deliver to fix the legacy infrastructure problems, while Whitehall is only slowly coming to terms with the scale of the climate change challenge.

Section 3.1 explains that the water companies have invested heavily in physical assets, but £150 billion of capital expenditure since privatisation has not been enough to sustain the infrastructure in this asset-heavy activity. Company investment strategies of patch and mend focused on like-for-like replacement, not prudent renewal and enhancement, so that in the 2020s there is a significant problem of under-investment. By the 2020s the stock of infrastructure assets was generally in poor condition, with companies going slow on the replacement of assets at the end of their useful life and neglecting upgrade and capacity increases in sewage treatment. System-wide publicly available information is limited, but we do know that, at 2002–22 replacement rates, it would take 100 years to replace the existing water mains and 350 years to replace the existing sewer mains.

Section 3.2 returns to the politics of water and shows how a trade association can confuse politicians and the public through misleading promises to fix problems while the industry continues to under-invest. Thus, Water UK – the trade association of the private companies – has a plan for the years up to 2050 which claims that the water companies will invest enough to deal with storm overflows. But this is an extend-and-pretend plan which postpones action and includes get-out clauses against a benchmark year when the number of spills was high because the year was unusually wet. The plan is also very narrowly focused since it does not address the issue of *E. coli* in treated sewage discharges, nor recognise how agricultural and road runoff, and pollution from abandoned metal mines, are as important as sewage discharges in causing poor water quality in river courses.

Section 3.3 approaches the politics of water from a different angle and considers the slow, dawning official realisation that long-run climate change is a huge challenge for the water industry. On official Met Office projections, substantially warmer, wetter winters and hotter, drier

Much more investment required

> summers will create a climate change problem which is as much about excess water as about drought. Successive official reports up to 2023 focused mainly on water shortages and downplayed the problem, so that it appeared manageable within existing frameworks. This was done by magical thinking about how fewer leaks and lower household water usage would reduce demand. A 2023 Defra report does represent the dawning realisation of the gravity of the problem, but at this stage it gives no indication of how the necessary catchment and national planning can be delivered.

3.1 Decades of patch and mend

Chapter 2 established that the privatised water industry is an asset-intensive activity under revenue constraint, without a sustainable business model. The consequence for investment in physical assets has been capital rationing, with decades of parsimonious patch and mend focused on like-for-like replacement, not prudent renewal and enhancement. This has resulted in a collection of assets that is inadequately maintained, with companies going slow on the replacement of assets at the end of their useful life and neglecting the need for upgraded facilities with new technologies and increased treatment capacity. Whatever the form of ownership, water is an activity which cannot avoid a level of ongoing capital expenditure that is high by the standards of other activities. But the problem of the privatised companies is that, although they had no option but to spend very large sums on the maintenance and replacement of physical assets, this was not enough to maintain and replace the assets, let alone enhance the capital stock. The economic regulator, Ofwat, was culpably negligent in failing to blow the whistle on the deteriorating condition of the industry's assets.

To make sense of the under-investment problem beyond abstract official categories that have little meaning for many readers, it is useful to begin concretely by describing the physical condition of Thames Water's infrastructure, which is as much of a problem as the company's debt burden. This is significant because Thames serves 16 million customers, more than one-quarter of the English population.[1] Relative to its peers, Thames is a high-investment company: between 1989 and 2023 it spent 52% of its sales revenue on capital expenditure, compared with an average 41% spent by the other nine regional water and sewerage companies. This investment spend excludes the £4.5 billion cost of the Thames Tideway super sewer, because that cost is carried by a separate financing and managing company, Bazalgette Tunnel Ltd, under PFI-type arrangements[2] described in Chapter 6.2. Despite Thames's higher than average investment, the outcome has been a decrepit estate with antiquated equipment, which is just about delivering for customers in service terms today, but with a risk of catastrophic failure tomorrow. This comes at a high cost to the environment, as was officially confirmed in 2025, with the Environment Agency for the first time publishing data showing the number and length of sewage spills from storm overflows: in the case of Thames, there were some 26,061 spills in 2024, lasting an average of 13 hours.[3] Thames itself acknowledges that in the past it was too dependent on 'sweating assets' and needs to urgently address its substantial 'asset health deficit'.[4]

Much more investment required

- The potential for disastrous failure was flagged in 2024 as a 'critical risk' for incoming Labour government ministers, who 'were warned by civil servants that the dire state of Thames' infrastructure is one of the most urgent problems facing the new government'.[5] The condition of the Coppermills water treatment works in Walthamstow, serving three million customers, was apparently 'particularly alarming'.
- The hard infrastructure problems related to tanks and pipes in treatment works are serious. At a July 2022 meeting, the CEO of Thames Water allegedly warned the Health and Safety Executive that Thames's infrastructure was a risk to public safety because of 'gas digesters that could explode in densely populated areas and near train lines; and of creaking trunk mains that ... could flood basements in minutes',[6] potentially drowning people living in basement flats.
- The problems are not simply related to the hard infrastructure in water and sewage treatment. Some of Thames's IT hardware is more than thirty years old and is kept going by cannibalising parts, while some essential systems are apparently running Lotus Notes software from the late 1980s and early 1990s.[7] The company is therefore at risk of IT systems failure and vulnerable to cybercriminal attack.
- The problem is not simply neglect of existing facilities but failure to install extra capacity to deal with population increase and the demand for new housing in the Thames Valley. Campaigners allege that three-quarters of the 106 wastewater treatment works in the Upper Thames do not have enough hydraulic capacity to meet the demands of the existing population.[8] And the Environment Agency has

blocked a major new housing development because it represents 'an unacceptable environmental risk' when Oxford does not have the wastewater treatment capacity.[9]

This description of the condition of Thames Water's infrastructure dramatises the problems of an industry which is teetering on a cliff edge, with equipment failure risks that are the consequence of years of under-investment. We should not assume that the rest of the industry is in significantly better condition, even though by the standard of other activities, companies have allocated a large proportion of their total spend to capital expenditure on physical assets, as we saw in Chapter 2. Over the whole period since privatisation in 1989 (in nominal terms and using their own classifications), the privatised companies have spent £146 billion on capital expenditure, with equal amounts spent on water supply and wastewater management.[10] Capital expenditure is expansively defined in the water industry and now includes work that in other industries would be classed as maintenance and classified as operating expenditure.[11] But even so, nearly £150 billion over thirty-five years is, by any measure, a large sum to spend on physical assets and, as we noted in Chapter 2, a huge cash drain on the water industry's resources. As Exhibit 3.1 shows, in this asset-intensive and labour-light activity, the water companies have consistently laid out more on capital expenditure than on operations, and the share of capital in total expenditure has risen slightly from around 53% between 1989 and the mid-2000s, up to 58% by the early 2020s.

What £150 billion bought was inadequate rates of equipment replacement and renewal, which are generally obscured because the relevant data on asset condition and capacity is not

Much more investment required

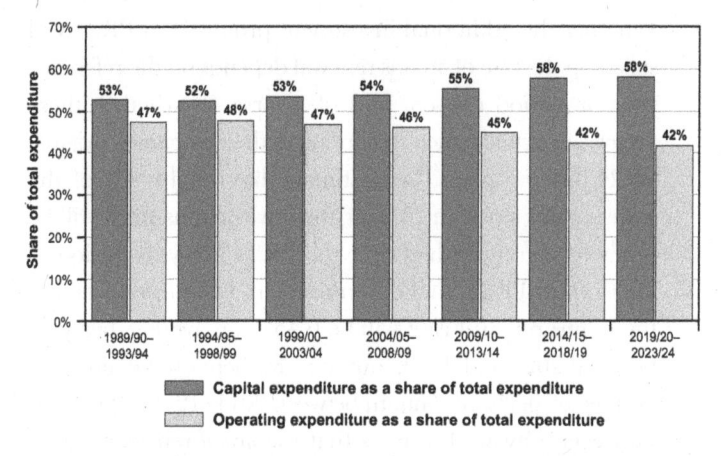

Exhibit 3.1 Capital expenditure and operating expenditure of water companies across price review periods, 1989/90–2023/24[12]

Capital expenditure as a share of total expenditure / Operating expenditure as a share of total expenditure:

- 1989/90–1993/94: 53% / 47%
- 1994/95–1998/99: 52% / 48%
- 1999/00–2003/04: 53% / 47%
- 2004/05–2008/09: 54% / 46%
- 2009/10–2013/14: 55% / 45%
- 2014/15–2018/19: 58% / 42%
- 2019/20–2023/24: 58% / 42%

in the public domain. However, rates of water and sewerage mains renewal can be calculated from Ofwat data, and these show how the companies have gone slow on end-of-life replacement investment. This evidence is highly relevant because two indicators of poor-quality infrastructure – 'mains bursts' and sewer collapses – are taken as proxies for asset health in Ofwat's 2023–24 Performance Review.[13]

- 18% of water mains were renewed or refurbished from 2000/01 to 2022/23 (see Exhibit 3.2), and, at this rate of roughly 1% per annum, it would take about a hundred years to renew the existing stock of mains water pipes. Some 6.1% of sewer mains were renewed or replaced from 2000/01 to 2022/23; and, at this rate of about 0.3% per annum, it would take about 350 years to renew the existing stock of sewer mains.[14]

Murky water

- Whether the additional investment promised in PR24 will speed up the rate of mains renewal depends on the reference year or period chosen, and of course on whether these targets are actually met. According to the companies' original PR24 business plans,[15] we estimate that roughly 4% of the water mains stock and 1% of the sewer mains stock will be renewed or refurbished over the PR24 five-year period to 2030. In its PR24 final determination, Ofwat predicts that the mains renewal rate will be 0.45% on average per year between 2025 and 2030,[16] an improvement on the historical average of 0.3% per annum between 2011/12 and 2023/24. However, Ofwat also notes that the *actual* renewal rate in 2020–25 was only 0.14% on average, compared with the 0.4% forecast in companies' business plans for PR19. There

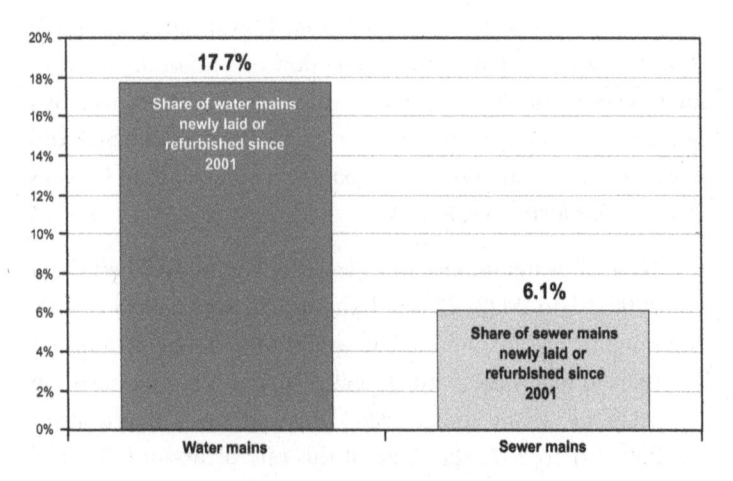

Exhibit 3.2 The proportion of newly installed or renovated water and sewer mains since 2001[17]

Much more investment required

must therefore be huge uncertainty about whether renewal targets will be achieved.

These renewal rates are culpably slow. There are approximately 280,000 km of water mains under the management of the ten major water and sewerage companies. Two-fifths of these are constructed in cast iron or ductile iron, 90% of which were installed before 1980, with a useful service life of 50–75 years. The sewer mains network is substantially longer at 576,300 km and is being renewed much more slowly. This slower rate of sewer mains renewal reflects the fact that gravity sewers deteriorate slowly rather than failing dramatically. Some water companies are going slow on 'rehabilitation' of their sewage pumping main stations which feed wastewater plants. Ofwat's work towards the PR24 draft determinations found that the overall 'rehabilitation' rate was 0.33% per annum from 2020 to 2024, but some companies had rates as low as 0.03%, which implies that one cycle of renewal would take more than 3,000 years.[18]

If renewal investment was being rationed, what Ofwat terms 'enhancement' investment or upgrade and adding new capacity has been starved. Hence we have an industry that by the 2020s has run out of hydraulic sewage treatment capacity at many points. As we have previously noted, a key study by Giakoumis and Voulvoulis in 2023 demonstrated that the acute problem of combined sewer overflows now reflects a 'chronic under capacity of the English wastewater system'.[19] Ofwat recognised the need for investment in upgrade and capacity increase for the first time when it introduced the concept of 'enhancement expenditure' in PR19.[20] The companies and Ofwat now recognise that more needs to be done. Therefore, enhancement expenditure on upgrade and capacity increase was projected to account for 52% of total capital expenditure in PR19 from 2020 to 2024 and is projected to account for 68% of capital expenditure in PR24 from 2025 to 2030.[21] However, it is far from certain that the companies have the capability to commission and manage a large increase in enhancement spend. According to Ofwat, under PR19 in the 2020–24 period, 'at a sectoral level water companies spent 16% less than their forecast enhancement allowance'.[22] Eight of the ten large water and sewerage companies underspent their relatively modest PR19 targets for enhancement expenditure in the five years up to 2024, including by 36% for United Utilities, by 35% for Thames, by 31% by Yorkshire and by 26% for Southern. This pattern of underspend is likely to continue in the next five years to 2030. Thus, headline investment promises about doing better should be treated with some caution.

But looking backwards, the damage has been done. Failure to prioritise enhancement in the 2000s and 2010s led to twenty years of patch-and-mend replacement. A key driver of this was the revenue constraint analysed in Chapter 2.2, which necessitated capital rationing. But patch-and-mend investment strategies have also been sanctioned – even encouraged – by negligent Ofwat regulation. The history of Ofwat's approach on this issue is a useful reminder of the limits of regulation when, as in this case, the regulator has a narrow brief and has historically been concerned to avoid conflict with companies and criticism from politicians and public. Thus, Ofwat focused on system performance and service-level objectives by setting key performance indicators (KPIs), and did not ask searching questions about patterns of investment and the underlying deteriorating condition and capacity of the companies' capital stock.

Much more investment required

In its annual company performance reviews, Ofwat has relied on backward-looking, short-term performance indicators. For example, in the 2023–24 Performance Review there were 12 objectives, each with corresponding KPIs: eight of the objectives covered system performance, including leakage, mains repairs, pollution incidents and unplanned outage, while four of the objectives concerned service level, including customer satisfaction and supply interruption.[23] This framework allows Ofwat to score companies as 'leading', 'average' and 'lagging behind', with this last group obliged to commit to performance improvement.[24] Thus, the annual performance reviews give water company managements a clear message that they must drive their equipment to meet system and service KPIs. If they meet enough KPIs they will escape attention, and no one will ask how this performance was achieved. The preoccupation with meeting KPIs can then distort investment priorities, as Ofwat admitted in its 2024 draft determination. One reason for declining rates of mains replacement was the company focus on repairs and pressure management, which 'have a greater short term impact on performance ... while this could improve performance in the short term (e.g. leakage, mains repairs), it could lead to a deterioration in asset health in the long term'.[25]

The Water Act of 2014 gave Ofwat an additional primary duty to further the long-term resilience of water and wastewater services, and Ofwat subsequently recognised that the water industry might have a problem with 'asset health and resilience', which could compromise delivery. But the regulator was not forensically curious about the nature and extent of that problem. Ofwat commissioned an independent report by the CH2M engineering consultancy, and accepted the consultants' 'nothing to see here, move along' reassurances in their 2017 report. CH2M recognised the fundamental problem that system performance or service-level indicators are not a good proxy for the condition of water infrastructure assets: 'it is possible, within limits, to provide an adequate level of service to customers even if individual assets are in poor health'.[26] CH2M was commissioned to undertake an interview-based study of senior water managers' perceptions of their own company's asset health,[27] so the resulting report was not a review of asset health using hard criteria, which would have disclosed accumulating problems. Instead, the consultants noted that the industry had a 'diversity of view' on the definition of asset health and reported senior management reassurances that in 2017 there were 'few immediately serious concerns regarding the impact of asset health on services', and that, therefore, a cliff edge of failure was not 'imminent'.[28] Looking back, this suggests a remarkable degree of complacency about the consequences of historical under-investment.

3.2 Water UK's fantasy plan

At this point in our argument, we return to the theme of the politics of water and the current political role of trade associations, both in general and in the specific case of water and sewerage. If regulators of privatised activities often inadvertently miss the point, trade associations in privatised activities can be relied upon to confuse the issue with special pleading. Privatisation came with economic promises about efficiency gains, management and competition (in the case of water, between companies, not for customers). The political outcome was the entrenchment of vested interests. Various public bodies – from nationalised

Much more investment required

corporations to local councils – with a blurred understanding of the public interest were replaced by corporates (representing investors), which had a sharply focused understanding of their own private interest. In a world of regulation and centrist two-party politics, privatisation greatly increased the importance of the trade association, that is, the organisation of corporate players in one activity who come together to lobby politicians and generally defend their collective private interest. The corporate players in a trade association have a vested interest in resisting changes that reduce profitability and promoting changes that increase profitability. They typically support their intra-elite lobbying with the production of what we call trade narratives about the public benefits of what the corporates have done and will do.[29] They will also be sensitive to public 'noise' and adjust the narrative to defend their record, sometimes emphasising that they are aligned with customer or wider public interests.

The classic trade narrative works by selecting evidence and listing the many good things that the public should be grateful for, while ignoring all the negatives. Thus, as we noted in Chapter 2.2, in the late 2010s the emphasis was on low and lower bills (while glossing over the bill increases of the 1990s which had funded meeting European Union directives). Tactics had to change by 2023–24, however, when the industry was under mounting public pressure from the media, civil society and politicians about storm overflows. In a shameless volte-face, Water UK now blamed Ofwat for the low bills that had caused under-investment. The second part of this repositioning was the promise that all would be well because the private companies had investment plans in hand to fix water's accumulated problems

Murky water

– hence the *National Storm Overflows Plan for England for 2025–50*, which Water UK published in March 2024 and presented on the Water UK home page as 'our world leading plan to remove 4 million spills from rivers and seas'. This plan merits sustained attention, not because it is a serious strategy that offers solutions, but because it is an extend-and-pretend plan whose misleading claims indicate the kinds of obfuscation that trade associations produce when they are under pressure.

The Water UK plan is presented as the industry's investment offer for what it would do under its own initiative to 'meet or exceed all government targets' on overflows and other issues.[30] A plan for the twenty-five-year period from 2025 to 2050 was produced by combining existing plans for the next five years with projections to 2050. The short-term plan for 2025–30 simply aggregated the English water companies' PR24 plans submitted to Ofwat. Beyond that, there was a long-term vision of what the companies *might* do in the subsequent twenty years from 2030 to 2050, which was not binding on individual companies nor the industry as a whole. 'Storm overflows' were foregrounded in the title and the text because the industry wants to be seen to be responding to public indignation about untreated sewage discharges, but the investment more broadly was also intended to reduce leakage by 28% over a decade and 'enable 10 new reservoirs to be built'. The implicit message on the Water UK website is that this is now a problem that will be sorted through the agency of the water companies. However, the plan is a fantasy whose effect is to compound the confusion of politicians, public and civil society through a farrago of misleading claims.

Extend-and-pretend is most obvious when it comes to dealing with storm overflows.

Much more investment required

- Water UK has an *extend* plan because it pushes the attainment of the overflow reduction targets into the distant future in 2050 and then backloads action to achieve these targets into later years. As Exhibit 3.3 shows, the Water UK plan draws an almost straight-line trajectory for reductions in discharges or 'spills' in 'priority overflows' to 2035 and expects further reductions of around four to five spills per overflow every five years to 2045, with more limited reduction after that. This projection of continuous improvement in overall spill reduction over twenty-five years to 2050 has the direct effect of allowing the industry to promise substantial improvement in the long term but avoid much of the action to achieve those large improvements in the medium term over the next five or even ten years. Modest and inadequate action over the period 2025–30 will deliver only 28% of the total planned reduction by 2050; and half of the overall improvement will not come until after 2035. Focusing on the *number* of spills also ignores their duration and the volumes discharged.
- Water UK has a *pretend* plan because it sets what appear to be exacting long-term targets for reducing the number of spills. On closer examination it is more complicated, because these targets come with undisclosed assumptions and explicit qualifications that give the industry excuses for missing them. In the long term, the vision target is just seven spills per overflow by 2050, down from 25 spills per overflow in 2025, and with 'no ecological impact from a storm overflow by 2045'.[31] But the ability to meet these targets depends on assumptions about climate change which are not specified in the published plan, so we do not know how it factors in Met Office predictions of wetter winters. In any case, there

Murky water

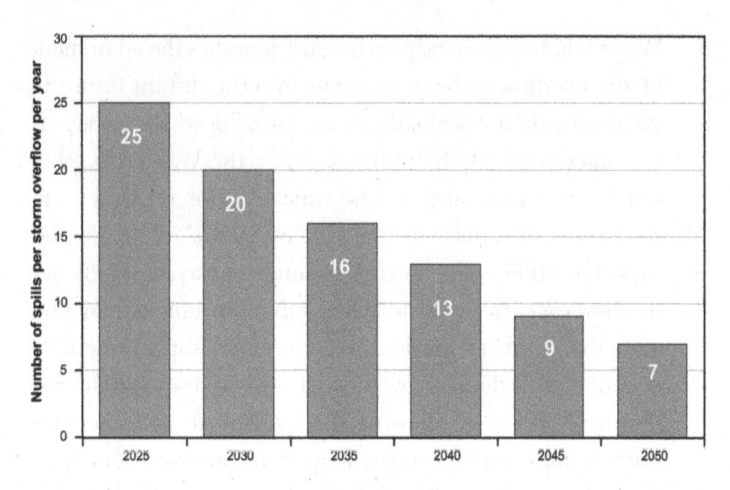

Exhibit 3.3 Water UK's expected annual frequency of anticipated spills from storm overflows in England, 2025–50[32]

will certainly be substantial year-on-year variation in rainfall, which allows Water UK to insert a get-out clause: 'While the overall trend will clearly be downwards, individual annual results will depend on how the weather has performed. All else being equal, there will be fewer spills in drier years and more spills in wetter years.'[33]

- The pretence is compounded by the way in which Water UK takes a recent wetter year as the benchmark against which long-run improvement will be measured. Water UK's claim is that the industry will invest £8.5 billion[34] to cut the number of overflow incidents (into rivers and on to beaches) by 'up to 140,000 each year by 2030 compared with 2020',[35] when more than 403,000 discharges were recorded. This is broadly in line with the Environment Agency's 2023 *Storm Overflows Discharge Reduction Plan*. But

Much more investment required

the reduction to around 260,000 spills is less impressive when we recall that the number of spills varies from year to year, and the Water UK baseline year of 2020 was a wet year. By contrast, in 2022 which was much drier there were 301,000 spills, around 25% fewer than in 2020.[36]

Water UK's plan for storm overflows is thus more a device for placating public indignation than a strategy for dealing rapidly and decisively with storm overflows, and its credibility is severely undermined by industry failures to meet earlier targets.

If we extend the analysis to other issues, Water UK's claims about what the rest of its plan will deliver come with a fair amount of spin. The one hard target is the claim that water supply will be improved as leakage is reduced by 28% 'over a decade' by 2035.[37] This ten-year target is broadly consistent with company PR24 business plan submissions for the five years to 2030. The simple average of the ten companies' promised five-year leakage reductions in PR24 is 19.2%.[38] But then the plan veers towards fantasy with the claim that 'the companies' proposals ... will enable 10 new reservoirs to be built'.[39] This is not a firm spades-in-the-ground promise about which reservoirs will be built where and when they will be completed. It is strictly true that the companies will 'propose' new reservoirs, sewerage treatment and water transfer schemes. But, as we show in Chapter 6, all projects with a total lifetime expenditure of more than £200 million will routinely be built on a PFI basis, because the debt liabilities of a major catch-up investment programme cannot sit on the wrecked balance sheets of the water companies. The 'will enable' phrase in the Water UK plan is therefore economical with the truth in a way that greatly exaggerates the agency of the water companies in solving our problems.

Murky water

Finally, Water UK's plan narrows the success metrics and problem definition in a way that reduces its responsibility for water quality. The overflow target is explicitly connected to a reduction in damaging phosphorus discharges, with river-quality monitors installed, but Water UK does not commit to any measurable improvement in river ecology so that the water-quality outcomes of its investment can be directly assessed. While Water UK's plan highlights investment in storm overflows which discharge untreated sewage, it ignores the industry's public health responsibility for *E. coli* in *treated* sewage discharges. A 2024 report by the National Engineering Policy Centre focused on the industry's contribution to poor water quality at bathing sites, specifically that 'the final treated effluent discharged continuously into water bodies still contains high numbers of faecal organisms'.[40] The ultraviolet disinfection systems installed 'at some coastal sites' can remove up to 99% of faecal coliform bacteria from sewage treatment plant effluent. But the water industry has no count of the number of unprotected sites, and no plan for fitting disinfection systems at least to priority sites over a specific timeframe.

If improvement in water quality is the broad objective, the Water UK plan does not admit that focusing on sewage discharges (treated and untreated) is necessary but not sufficient from a water management point of view. There are other major contributors to poor water quality, including agricultural runoff, road runoff and pollution from abandoned metal mines. All these sources of pollution now have much the same status as storm overflows a decade ago: they are known unknowns which are not being systematically measured, regulated or remedied by any government agency. What we do know is that the water quality in our rivers is in a dismal state. According to the Environment Agency,

Much more investment required

under criteria established by the EU Water Framework Directive, in 2023 only 14% of English rivers met 'good ecological status',[41] no river met 'good' chemical status, and only two stretches of river in England and Wales had 'bathing water' status, both of which were rated as 'in poor health'.[42]

One of the problems here is limited knowledge. We have traditionally relied on occasional laboratory tests of the microbiological quality of treated effluent, because reliable, real-time monitoring of water quality is only now becoming practical and affordable. These knowledge problems are almost inevitably politically multiplied when there is dispute about causality, as there is about the relative contribution of sewage discharges and agricultural runoff to the alarming deterioration of the aqua systems in English and Welsh rivers. At a House of Commons Environmental Audit Committee inquiry in 2021–22, expert witnesses and environmental NGOs generally agreed that agriculture did play a major part, but representatives from the water industry and from agriculture competed in an unedifying way to deny responsibility and blame each other for poor water quality in rivers.[43]

The agricultural problem is caused by rainfall runoff after catchment overloading of nitrogen and phosphorus residues from fertilisers and manure, especially from intensive livestock and poultry farming. For example, in the Wye Valley, planning laissez-faire has allowed the building of more than a thousand poultry sheds,[44] which contained an estimated total of more than 20 million farmed birds by 2020. The result is eutrophication, whose visible consequences are algal blooms that deplete oxygen levels, with resultant damage to river ecosystems. In the Wye Valley this has led to a claimed 90% loss in the protected Ranunculus bed, and severe declines in fish species such as salmon and white claw crayfish.[45] The damage is catastrophic and, although environmental campaign groups such as River Action have highlighted the scale and nature of the challenge,[46] a blame game between water and farming industry representatives covers slow progress towards investing in decisive remedial action by both parties, with neither taking responsibility for fixing its own part of the problem.

The water companies are not planning to invest seriously and adequately in the problems they are clearly responsible for, such as their treated and untreated sewage discharges. They appear content to let other problems such as agricultural runoff (which they are not directly responsible for) have a low profile. This makes it less likely that they will be addressed in any other organisation's plan; moreover, it is an obstacle to necessary attempts to develop holistic approaches to water quality. All this secures the legitimacy of the water industry by encouraging the illusion that its plan to tackle storm overflows will solve the problem of water quality. Perhaps the most serious of these low-profile problems is road runoff, which has pervasive implications because of the way that waterways are used as a sink for the runoff from tarmac surfaces. There are 18,000 known outfalls on motorways and trunk roads,[47] with countless others on lesser roads which often reach waterways via the conduit of surface water sewers.

Rain washes a toxic mix of tyre and brake particles, fuel, metals, road surface and herbicide through soakaways and outfalls into waterways, without any monitoring or treatment, though mitigation technologies and devices are readily available.[48] The harm is poorly understood, but road runoff typically includes locally persistent, toxic compounds such as polyaromatic hydrocarbons (PAH), while vehicle tyres are supposedly the largest terrestrial source of microplastics in the oceans.[49] All this is directly the responsibility of the Environment Agency, National Highways and local authorities. But in an interconnected, complex system it is also relevant to the water companies. From a system point of view, the Chartered Institute of Water and Environmental Management argues that it could often be cheaper and easier to reduce levels of some pollutants such as benzopyrene in waterways by treating road runoff, rather than adding more treatment of wastewater discharges.[50] Addressing this issue requires collaboration between water companies, highways agencies and local authorities; as private-interest entities, the water companies have little incentive to initiate such dialogue.

Finally, if water quality is the central issue, beyond agricultural and road runoffs, there is the problem of heavy metal pollution from long-abandoned mines, whose workings and spoil tips contain large concentrations of metals such as cadmium, iron, lead and zinc. As long ago as 2008, the Environment Agency admitted that 9% of rivers in England and Wales were at risk of failing to achieve targets of good chemical and ecological status because of abandoned mines.[51] In January 2023 the UK Parliament approved a legally binding target to halve the length of rivers polluted by abandoned metal mines by 2038. However, the *Financial Times* reported in 2024 that the Environment Agency and Defra were currently operating just four main water treatment schemes, and argued that the risk of lead getting into the food chain had been seriously underestimated when the UK has more than 6,000 abandoned lead mines.[52]

The simple and dismal conclusion is that Water UK's storm overflows plan shows how trade narratives spread misinformation which can easily confuse political elites and the general public into thinking that problems are being addressed when they are not. If we turn from the Water UK plan to the companies' submissions to Ofwat for the PR24 price determinations, the submissions propose an increase in investment in physical assets from around £4 billion a year in PR19[53] to £10 billion a year by 2029 (in 2023 prices).[54] This £50 billion of investment in physical assets in PR24 is not enough to significantly improve the quality of treated and untreated discharges. And none of this addresses the larger issue about how water companies' investment fits into systemwide needs and priorities for pollution mitigation, in domains that are outside the direct control of the companies. Nature-based solutions are completely marginalised when 75% of the PR24 interventions on sewer outflows are in fact 'traditionally engineered' or 'smart sewer projects'.[55] The challenges are large, and the Water UK plan offers a response which is inadequate, even without considering the long-term impact of climate change, which will have a significant impact on temperature and rainfall in ways that need to be addressed within the management of the water system. We turn to this in the next section.

3.3 Dawning realism about climate adaptation

We begin by briefly summarising current predictions about the extent of climate change by 2100 and the resulting water management challenges. Chapter 1.3 outlined how climate change in the UK has already brought significant changes. With mitigation to reduce emissions failing to meet targets, these processes will continue, so that, by 2100, on official Met Office projections, the UK will probably be 4 degrees warmer. This creates a massive adaptation challenge. Substantially warmer,

Much more investment required

wetter winters with more intense rainfall,[56] and hotter, drier summers, will interact with the west-to-east rainfall gradient in the UK to create a large and inescapable dual problem regarding the management of both drought and excess water. In the second half of this section, we consider the official response. Up to 2023 there was a head-in-the-sand period when expert and technocratic elites downplayed the nature and scale of the challenge by magical thinking about leak reduction and reduced personal consumption. In this way they met the political classes' need to portray climate change adaptation as manageable within the existing water industry frameworks. Recent official reports have recognised the nature and scale of the adaptation. This requires substantial new infrastructures for water transfer and storage, and a new approach to integrated water management involving nature-based as well as steel and concrete solutions. But these reports do not, and politically cannot within the existing Overton window, specify how to get there.

Long-term climate changes are difficult to predict precisely because there are so many variables and relations involved in jointly determining outcomes. Consequently, the UK Met Office has a total of 20 scenarios, from low-emission best-case scenarios to high-emission worst-case scenarios. It is increasingly clear that collective action at the global level is failing to mitigate emissions within Paris Agreement limits, and we should now at national level prepare for worst-case climate scenarios in making adaptation plans. This is what the Met Office itself currently recommends. The worst-case approach of UK climate scientists fits with the longstanding practice of the engineering profession when making design decisions on critical infrastructure in systems such as water and sewerage. Engineers from Joseph Bazalgette in the nineteenth century to Deb Chachra in the 2020s have recommended redundancy and over-specification for inbuilt system resilience in an uncertain future.[57] The empirics below are therefore all based on current Met Office no-mitigation, worst-case scenarios.

As we noted in Chapter 1.3, in England and Wales there is a rainfall gradient between the wetter north-west and the drier south-east. The water management challenge is that under a worst-case climate change scenario, the north-west will by 2100 become much wetter in the winter and the south-east will become much drier in the summer. This is clear from a comparison of the baseline period of 1981–2000 against Met Office worst-case predictions for 2080–99.[58] Without mitigation, global emissions and temperatures will continue to rise, leading to a global temperature rise of +5°C by end of the century.[59]

- In the baseline period of 1981–2000, north-west England and North Wales had 385 mm of rainfall in winter, which is 6% more than the national average of 337 mm, 80% more than the south-east including London, and double that in East Anglia. In 2080–99, winter rainfall in north-west England and North Wales is projected to rise by some 20% up to 466 mm (Exhibit 3.4a).
- In the baseline period of 1981–2000, East Anglia, London and south-east England had roughly 150 mm of summer rainfall, which is 30% less than the national average of 226 mm and 40% less than the 263 mm in north-west England and North Wales. In 2080–99, summer rainfall is projected to decrease by 38% to just 90–97 mm in these areas (Exhibit 3.4b).
- The effects of less summer and more winter rainfall will be magnified by the increase in average temperatures.

Much more investment required

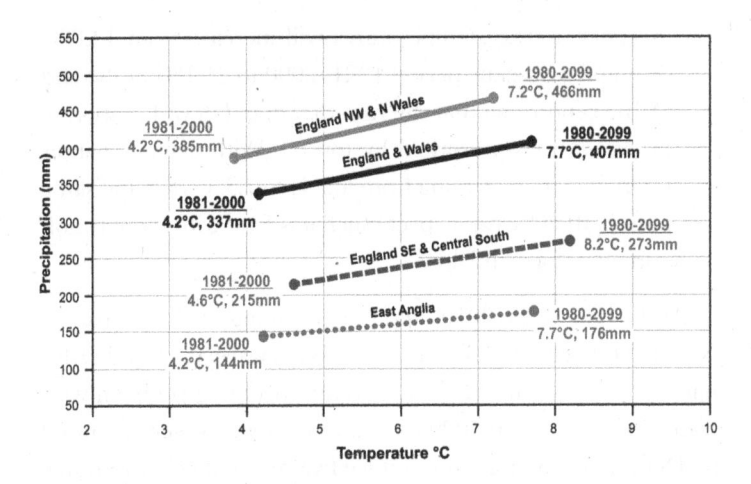

Exhibit 3.4a Current and anticipated seasonal average temperatures and precipitation levels, 1981–2000 and 2080–99, winter[60]

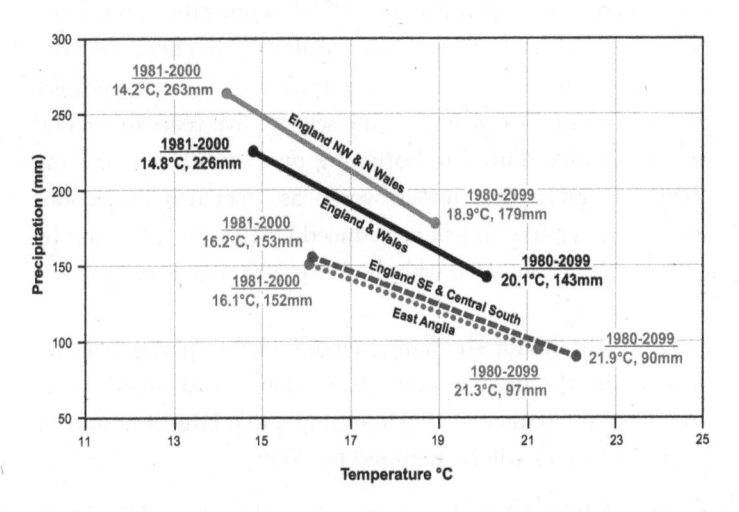

Exhibit 3.4b Current and anticipated seasonal average temperatures and precipitation levels, 1981–2000 and 2080–99, summer[61]

Murky water

Summer average temperatures will increase from 14.8°C during the baseline period 1981–2000 to 21.1°C at the end of the century, so that the rain that does fall will evaporate faster, with less absorbed by topsoil. With average winter temperatures increasing from 4.2°C to 7.7°C over the same period, the UK can expect much less frost with more runoff and waterlogged fields.

At the end of the 2010s and the beginning of the 2020s, the National Infrastructure Commission (NIC) warned[62] – and Defra had to admit[63] – that the UK had an upcoming problem with water shortage. So in 2020 the water companies were required by Defra to collaborate to produce five regional water resource plans in line with their company-level Water Resources Management Plans (WRMPs) to explain how they would address this shortage. The extent of the problem was reduced by concentrating on estimating water shortage up to 2050, arguing that 'projections beyond 2050 carry increasing uncertainty'.[64] In effect, the NIC and Defra reports considered the effects of climate change over the next twenty-five years, not the seventy-five years to the end of the century. This foreshortening dramatically reduces the challenge, given that under worst-case scenario projections summer precipitation will be reduced by 19% by 2050 and by 37% by 2100, compared with the 1981–2000 baseline (as shown in Exhibit 3.5).

Even with this foreshortening, climate change plus population growth and the need to curb abstraction – as outlined in the appendix to Chapter 5 – together imply that large amounts of additional water will be required by 2050.

- The NIC published the first high-level analysis, *Preparing for a Drier Future*,[65] in 2018, and estimated that an additional

Much more investment required

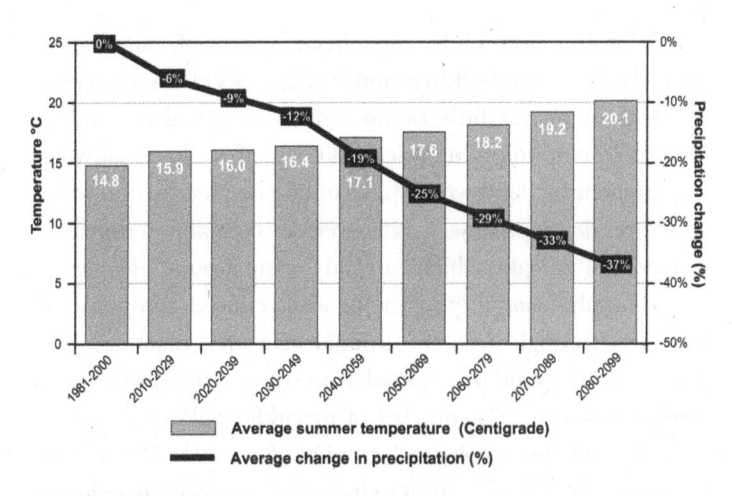

Exhibit 3.5 UK Climate Change Projections (UKCP18) for England and Wales: average summer temperature and changes in precipitation[66]

3,500–4,000 million litres per day (mlpd) would be required for England by 2059, a 25–27% increase on current water consumption of 14,000 mlpd.

- In 2020 the Environment Agency produced the first *National Framework for Water Resources*,[67] and estimated that the extra requirement was 3,400 additional mlpd, just below the NIC estimate.[68] It added in a throwaway that 'something in the region' of 5,500–6,000 mlpd of additional water supply would be required by 2100.[69]

The additional requirement was later revised upwards by the Environment Agency, so that the official target of government and industry in 2023 was for an additional 4,800 mlpd by 2050[70] (and presumably higher again by 2100). The water industry, even with the planned large-scale PFI projects and without the five-year price review time horizon, cannot conceivably increase supply on this scale under the current system of charging for water, because, on the Environment Agency's own estimates in 2020, if there was little or no reduction in leakage and/or household consumption, at least 80% of the additional water requirement by 2050 would have to be provided by new infrastructure developments.[71] However, if companies somehow reduced water leaks, which currently waste around one-fifth of water supply,[72] *and* if households also reduced consumption, including through stringent drought measures, much less new infrastructure would be required. According to the Environment Agency, between 50% and 70% of the additional water required by 2050 could be covered by a combination of 21–24% leakage reduction and 27–47% 'water efficiency' (significantly reducing consumption from the present 140 litres per person/per day).[73]

These magical assumptions about demand reduction are carried over into the WRMPs prepared by the companies.[74] The Environment Agency's 2024 *Revised Draft Regional and Water Resources Management Plans*[75] can therefore envisage 'almost two thirds of the [additional] water needed in 2050 coming from reductions in demand', of which 48% will come from 'using water more efficiently and metering', while 17% will come from 'reducing leakage'. The key driver of demand reduction in company plans will be a rollout of smart meters. Currently around 60% of households in England (including empty properties) are metered, but only an estimated 13% of households have smart meters that allow monitoring of water usage in real time.[76] In their 2024 WRMPs, water companies envisage around 48% smart metering of households by 2030, rising to 73% by 2040 and 76% by 2050.[77] Based on the impact of smart meters on household energy use, a 2024 consultancy report for Ofwat sets out expectations of reduced household water consumption

Much more investment required

of up to 20% by 2050, which is consistent with the Environment Agency scenario discussed above.[78]

These assumptions about reduced leakage and household consumption in official reports and company plans appear reassuring because they reduce the requirement for new supply capacity. The effect is to make the future problem of organising water supply under climate change appear manageable, so that Westminster and Whitehall need not panic. The difficulty is that the magical assumptions about leakage reduction and household consumption are completely implausible. And it is worth exposing the implausibility of these assumptions because this highlights how the official technocratic mind in Defra or the Environment Agency is capable of generating reassuring confusions, even without the financial interests that direct a trade association.

On leakage reduction, it is assumed that the future will be very unlike the past, even though the drivers of leakage reduction will not change. The record of the last two decades is a very slow reduction in leakage losses. Between 2005–07 and 2021–23, leakage was reduced by 14% from 3,580 to 3,089 million litres per day.[79] Exhibit 3.6 shows the scale of this reduction on a per-person per-day basis. The government and industry now target a far more ambitious additional 30–50% reduction in leakages over the next twenty-five years (from the 2018–19 baseline).[80] An improvement in the rate of leakage reduction might be plausible if the industry was planning to dramatically increase the rate at which it is replacing post-1945 cast iron mains pipes that have reached the end of their useful life. But as we noted earlier in this chapter, under PR24 plans the companies are unlikely to deliver significantly increased water mains renewal rates.

Murky water

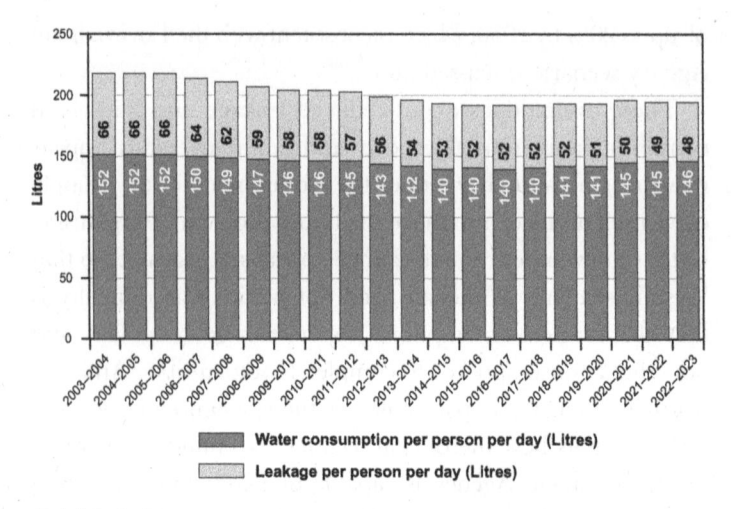

Exhibit 3.6 Daily average water consumption and leakage per person in England, 2003/04–2022/23[81]

The official projections of reduced household demand rest on the implausible assumption that smart meters will deliver a significant reduction in household consumption. The target of equipping 70% of households with smart water meters by 2050 is realistic, given that by September 2024, 37 million electricity meters or 65% of the total were smart.[82] However, 60% of household properties are already water-metered at present in England, and the switch from dumb to smart meters will most likely result in a negligible reduction in consumption. The NIC estimates that conventional metering reduces demand by around 15%, and smart meters are expected to only add a further 2% reduction.[83] This estimate is confirmed by independent academic studies (involving experimental change and large samples) which confirm very modest long-term reductions in household demand after the introduction of smart meters.[84]

Much more investment required

Water consumption has been more or less steady at an average of 140–150 litres per person per day over the past two decades, as indicated in Exhibit 3.6. It is difficult to reduce consumption from this plateau because it is structurally embedded in cultural practices such as the daily shower, and in household technologies such as front-loading washing machines and dishwashers. It is quite unrealistic to expect that this can be reduced to 110 litres per person per day by 2050, as envisioned by Defra and Ofwat,[85] without explicit rationing or drought measures (or punitive pricing). Lower usage of 85 litres per person per day was possible in the 1960s,[86] when water use habits were different. We can realistically expect some reductions in household water use as appliances such as washing machines and dishwashers become more efficient. But a major shift is unrealistic, considering also that technology often promises more than it delivers, as the case of dual-flush toilets shows. Dual-flush toilets were intended to reduce water use, but are now 'wasting more water than they save',[87] and they are the primary cause of toilet leakage of 400 million litres a day.

Against this background, the good news is that by 2023–24 the Environment Agency and Defra were beginning to recognise and engage with the nature, extent and intractability of the upcoming water management problem. The Environment Agency's 2024 flood risk report was a landmark which broadens the problem definition to consider water excess and puts predictions on a more realistic basis. The Defra *Plan for Water* in 2023 for the first time had a vision of a coherent catchment-based approach to water shortage and excess.

The Environment Agency's *National Assessment of Flood and Coastal Erosion Risk in 2024*[88] put the issue of water excess in a period of climate change on to the agenda. For the first time in the UK, this report factored climate change projections into its assessment of flood risk.[89] The outcome is the highlighting of the large number of houses at risk of flooding, as well as the potential for large-scale economic disruption. In short, one in four properties in England could be at risk of flooding from river, sea or surface water by mid-century.[90] Of these, the largest number are at risk of shallow flooding by surface water (not rivers or the sea), caused by rainfall beyond the absorptive capacity of the ground and the clearance capacity of the drains. Significantly, the Environment Agency projects a 30% increase in properties at risk of surface water flooding, from 4.6 to 6.1 million by 2050.[91] The more general potential for economic disruption is also expected to be considerable. More than half of grade one agricultural land is currently at risk of flooding from rivers and the sea, and around half the rail and road network will be at risk of flooding by mid-century.[92]

Defra's *Plan for Water* report of 2023 represents the official mind at its most imaginative, because in many ways it has a radical and coherent vision of where we want to get to.[93] The ambition is to transform the management of the water system through 'an integrated approach across a whole catchment' to deliver 'a healthy water environment and a sustainable supply'.[94] The ministerial foreword promises a 'coordinated and collaborative' approach, 'using both nature-based solutions and investment in infrastructure involving communities, water companies and businesses'.[95] The vision clearly now includes water excess as well as water shortage, and the 2023 plan for water recommends a catchment-focused approach to the adaptation of urban and rural environments, listing what is technically necessary to deal with water excess.[96] For example, in urban areas a key recommendation is to equip all new developments with sustainable

Much more investment required

drainage systems,[97] so that more rainwater is retained where it falls and combined sewer networks are not overwhelmed, though retrofitting of existing developments is also likely to be required.

If this report represents a large step forward, disappointingly there is very little on tree planting and other methods of water retention in rural areas to reduce runoffs into lowland areas,[98] while some aspects of the water excess problem such as the waterlogging of farmland in winter are notably absent. More importantly, the report fails to confront the question of how to realise the vision and address delivery challenges in terms of both technical solutions and collaborations. There are to be more catchment 'action plans', 'groups' and 'partnerships' which have been piloted on a voluntary, small scale.[99] But there is nothing about the systems of rewards and punishments which, as part of a coordinated planning and management system, would ensure that effective collaboration happens at scale. Defra's treatment of 'water supply' still focuses too much on water shortage, not water excess and water quality. On water shortage, the report does not break with old ways of thinking, and Defra's role is to accelerate the infrastructure investment which the industry already has in hand. Future water shortage is once again presented as a problem that can be made manageable with demand reduction, if companies are set 'ambitious targets' for leak reduction, and through new product standards, such as for showers and toilets.[100] Much action on water quality is simply postponed: Defra is doing no more than 'considering actions' to reduce road runoff and promising 'consultation' on sustainable drainage systems, but only in 'new developments'.

The implications are very clear. Beyond the challenge of fixing the backlog of infrastructure renewal highlighted earlier

Murky water

in the chapter, we need to understand that the problem of water management up to the end of the century is not a narrow technical issue for the existing water industry, requiring finance and civil engineering expertise. The greening of urban and rural environments in our century of climate change is a *political* problem requiring coordinated action by multiple actors to deliver agreed objectives. Achieving these will require the disruption of comfortable consensus thinking by political elites and the technocrats who serve them. The water company (public or private) as we know it is, in broad historical perspective, an invention of recent date. It is now nearing its end, in relation to how management of the water system needs to change. With the challenge of water management under climate change, we will have to invent new institutions, just as the Victorians did. We return to these important issues in Chapter 6. Meanwhile, whether we look at the self-serving obfuscations of Water UK or the slow official recognition of the gravity of the climate change adaptation problem, the message of this chapter is that with water we are in politics, and we will need in Chapter 5 to understand how politics is increasingly driven by issue management and confused by regulation.

Chapter 4

From water poverty to water justice

Introduction

This chapter argues for a key shift in the way we think about how households pay for water. The argument is to move away from the notion of water poverty, classically tackled by a social tariff offering price rebates for some low-income households, and move towards the idea of water justice, tackled by progressive charging for water based on household income. This would mean not only that lower-income households pay less, but that high-income households pay their fair share. In this way we recognise the collective public interest in a water system that is affordable and can sustain itself.

Section 4.1 on how households pay for water – based on meters or rateable value – takes us further into the history of water, money and power. The problem of the charging system has been understood through the lens of water poverty in low-income households. Led by the

Consumer Council for Water, the social lobby's preferred solution was and is price rebates through social tariffs for low-income households and those with higher water needs. Water companies were allowed to create a postcode lottery through individual social tariff schemes and at the same time kept average bills low, with the unintended consequence of revenue constraint and business model problems. When it was clear that higher bills were coming, in 2024 Water UK finally accepted that a unified national social tariff scheme was inevitable. But, as section 4.2 shows, the water companies have ensured that the cost of price rebates falls mainly on other households that still pay the full price, not on water company shareholders.

Section 4.2 explores the regressive effects of the existing household charging system in England and Wales by arranging households into decile groups and calculating spend on water as a share of their total expenditure. On a household basis, the empirics show that low-income deciles pay proportionately more for their water and high-income households pay less; on a per-person basis, the poor pay absolutely more. The regressivity problem is as much about a system where the rich pay too little for their water as it is about the poor paying too much. Price rebates through social tariffs therefore only address one half of the regressivity problem, which is about the burden on poorer households. Moreover, social tariff schemes solve one problem at the expense of creating new problems. As with other means-tested benefits, there will be problems related to take-up rates among those who are eligible and cliff-edge inequalities where entitlement to social tariffs begins and ends.

Section 4.3 brings the good news on social justice. Alternative forms of household charging with progressive outcomes are both possible and practical, if bills for water are varied according to household income. This is empirically demonstrated in two new simulations. The first models a flat rate, with all households paying the same percentage. The second models a progressive charge with tapering payment rates, so that, at the bottom of the distribution, households pay almost nothing for water while households at the top of the distribution pay more (though this is still a relatively small amount in relation to their overall income). Both simulations deliver more equitable outcomes and would also allow the industry to raise more revenue and address its business model problems. Household charging related to income is sensible, given the large differences in ability to pay, and it is politically feasible when our simulations show that there would be many more winners than losers. Critically, this approach also recognises that water and sewerage should be considered as a public service rather than a private consumption good.

4.1 Charging through the lens of water poverty

Household charging for water is currently a political mess with an instructive history. Like much else in water, this history is about narrow problem definitions and power relations, as the narrowly focused, well-meaning concern of the social lobby on water poverty was frustrated by the self-interested resistance of the water companies. The outcome was twenty wasted years when the problem of water poverty was not tackled by adopting a national social tariff which would have rebated the price paid by low-income households. Under a regressive system of charging for water, high-income households were not paying their fair share but, given the preoccupation with poverty, this never became a political issue. The water companies were allowed to set up individual social tariff schemes and create a postcode lottery whose effects were moderated by keeping average bills low. The unintended consequence was to reinforce revenue constraint and the business model problem. With general recognition in 2023–24 that higher bills were unavoidable, everybody including Water UK finally accepted the need for a national social tariff. But the ramshackle system of charging households according to metered consumption or rateable value has never been brought into focus.

This charging system was understood after privatisation through the lens of 'water poverty'. The concept passed into widespread use in the 1990s and has since been understood as 'the condition in which households struggle to afford their water bill and suffer economically and socially because of it'.[1] Underpinning the concept of poverty is the idea of a right to water, as reflected in Goal 6 of the 17 UN Sustainable Development Goals, which is to 'ensure availability and sustainable management of water and sanitation for all'.[2] Attaining this goal is generally understood to require physical access to infrastructure systems in the Global South, and affordable bills in the Global North. The concept of water poverty has considerable moral appeal,

From water poverty to water justice

given the necessity of access to clean water, but it does narrow the field of the visible. From a broader redistributive perspective, the issue is not only whether low-income households in the bottom third of the income distribution are paying more than they can afford, but also whether high-income households in the top third of the income distribution could afford to pay more to support the foundational infrastructure. Thinking in this way also recognises that water and sewerage is a public service – not a private consumption good – that should deliver social, economic and environmental benefits to all.

In the UK, 99% of households are connected to mains water[3] so that physical access is mostly not compromised by the limited extent of the UK network, though mains sewerage is less comprehensive. Water poverty surfaced as a problem in the 1990s after privatisation, initially via concerns about termination of access through disconnection. The newly privatised water companies were more willing than the public authorities they replaced to disconnect households that failed to pay their bills, which were at that time rising to cover the costs of clean-up mandated by EU directives. In 1991–92 disconnections peaked at 21,000 households, but by 1994 many more were at risk, with at least two million households in debt to their water company.[4] The result was civil society outrage about disconnections led by figures such as the head of the British Medical Association. The Labour government responded in 1999 with the Water Industry Act, which made disconnections illegal. At this point, water poverty ceased to be a matter of high-profile outrage about disconnection and became a rumbling matter of concern about affordability. The problem of customer arrears persisted because it was an index of affordability problems.

Murky water

Issues related to households who could not easily pay their water bill were recognised and addressed in two stages.

- In 1999 'vulnerable groups' regulations proposed assistance for households with high water consumption. The outcome was a national WaterSure scheme in England and separately in Wales. These assisted metered households with three or more children or with specified medical conditions, such as psoriasis, by capping their bills at the average for the area so that they were not penalised for their higher needs.[5]
- A 2010 Act licensed social tariff schemes to reach broader groups of beneficiaries facing affordability problems. European social tariff schemes are typically national, primarily offer price rebates, and have clear eligibility conditions based on income or benefits status. Bizarrely, in 2012 Defra allowed each privatised water company to develop its own social tariff scheme 'which meet(s) the needs of their consumers'.[6] Consequently, eligibility conditions have never been standardised across regions and the benefits provided are uneven. Defra was not prepared to override the companies who objected to a national scheme because they feared it would involve some element of cross-subsidy and cash transfer between different companies, given the differences in income levels between regions.

Affordability problems persisted and in recent years have worsened following the cost-of-living crisis. The standard pre-crisis quantitative criterion for water poverty was that households spend more than 3% or 5% of their disposable income (after housing costs) on water. In 2019, before the pandemic and the cost-of-living crisis, a study in England and Wales found that at the 5% threshold, almost 1.5 million (6.5% of the total) were in water poverty, and, at the 3% threshold, over 4 million (17.9% of all households) were in water poverty.[7] The position by 2023 and 2024 was different and worse. Although water charges had not increased in real terms, as noted in Chapter 2.2, the rising costs of on-market essentials such as energy and food had squeezed budgets and consequently made it more difficult to pay for water. A May 2023 Savanta opinion poll for Ofwat found that 23% of respondents had struggled to pay their water bill in the previous twelve months.[8] In March 2024 Ofwat found that 2.5 million households (8.2% of all households) were in arrears and the average debt was a massive £882 (around two years' worth of bills, on average).[9] In November 2024 a Consumer Council for Water (CCW) survey found that 40% of customers said they would find it difficult to afford the bill increases proposed by the water companies for 2025–30.[10]

The WaterSure schemes and company-based social tariff schemes have both, in different ways, failed to address the problems of affordability and debt. In WaterSure, the eligibility conditions are so restrictive that the numbers assisted have always been small and have increased only slowly. In 2010–11 just 43,000 households benefited from WaterSure,[11] and by March 2023 the number had increased to just over 220,000 households.[12] The company-based social tariff schemes have a take-up problem, partly because of their confusing variety and general reluctance to offer price rebates. The only support universally offered by all schemes in 2023 was 'payment breaks'. Other forms of support were company-specific so that, for example, 10 of the 17 regulated water companies had a 'financial hardship fund', while seven (including Dŵr Cymru, Northumbrian and Wessex) did not.[13] The result is what the House of Lords Industry and Regulators Committee in 2023 described as a 'postcode lottery' for access to support, in a situation where three-quarters of customers do not know that water companies will offer help with bills.[14] An estimated 5.7 million households are eligible for water company schemes; 'however many have not claimed and are missing out on savings of around £160 per year'.[15] In early 2023 Water UK estimated that just 1.1 million households (out of nearly 25 million in England and Wales) were receiving some form of benefit under the various company schemes.[16]

The history of the WaterSure and company-specific social tariff schemes is of missed industry targets for numbers assisted in the 2010s, and then of new plans and rapid increases in the numbers assisted after the cost-of-living crisis in 2023–24. In PR14 (covering water charges for 2015–20), Ofwat set the industry a target of assisting an additional one million low-income households, though the outturn was just 140,000.[17] The shortfall was covered over by a ramping up of the industry's rhetoric. In 2019 the industry made 'a public interest commitment' to 'make bills affordable as a minimum for all households with water and sewage bills more than five per cent of their disposable income by 2030, and also to develop a strategy to end water poverty'.[18] This target will not be met, though by 2023–24 the numbers assisted had increased fairly rapidly due to mounting cost-of-living pressures. In October 2023 Water UK announced a target of 3.2 million consumers to be assisted by 2030,[19] and in February 2024 it claimed, rather ambiguously, that two million households 'now receive some sort of financial support with their bills'.[20] The latest, most authoritative statement of bill rebate targets for 2025–30 comes in Ofwat's PR24 final determination. The claim is that 'companies now plan to support 9% of customers with reduced charges … compared to 4% in 2020–5'.[21] This is better than nothing, but remains an inadequate response when household incomes have been squeezed by the cost-of-living crisis and water bills are now also rising.

The failure of water poverty assistance has had consequences at company and system level.

- At company level, the cost of price rebates and bad debts is being covered by surcharges on the bills of other households. In 2011 the government introduced an 'enhanced' WaterSure scheme, 'with the cost met by the government'.[22] But by 2023, the CCW found that WaterSure 'is now primarily funded through cross-subsidy from other customers' bills' and was currently adding £2 to £3 per year.[23] The much larger and increasing cost of bad debt is being covered in the same way. By 2020 some £3.5 billion of bad debt was being covered by a surcharge of around £20 on annual household bills.[24]
- At system level throughout the 2010s, Ofwat was under increasing pressure to manage affordability problems by keeping overall bills low. With water poverty assistance stalled, the upward trend of bills in the 1990s could not continue. Hence in its PR14 final determination, Ofwat promised that average bills would be 5% lower in 2019 and in PR19 it promised 12% lower bills. After 2022, media and political concern with water poverty and the circumstances of households at the bottom of the income distribution was joined by a heightened concern with the average bill and the plight of ordinary households in the middle of the income distribution. This reflects cost-of-living pressures squeezing middle-income households in the 2020s, and the impact of water charges on a much larger group of consumers and voters.

Since 2020 the CCW has pressed the case for a single social tariff with a central pot of funding. Through the water poverty lens, this was a common-sense recognition within a water poverty framework that the industry schemes had failed. Others quickly recognised the inevitability of a single social tariff in 2023–24 when the companies submitted plans for PR24 involving headline average price increases of 'up to 70%'. These were moderated by Ofwat's final determination at the end of 2024, which settled on an average increase of 36% in 2029–30 compared with 2024–25.[25] In July 2023, when substantial PR24 increases were first mooted, the CCW protested that this would push a further 1.1 million households into water poverty by the 5% standard.[26] The CCW was then joined by charities and other campaigners, as well as the Social Market Foundation, in calling for unified affordability support to tackle the current postcode lottery for households needing financial support.[27] Finally, Water UK, which had long defended separate company schemes, recognised the inevitability of change, and by early 2025 was running ahead of Ofwat and the UK government in lending its support for a *national* social tariff.[28]

In all of this, the basis for charging for household consumption of water via direct billing in England and Wales has been taken for granted, though it is by no means universal. In the Republic of Ireland, for example, general taxation pays for water and wastewater services, no direct charge is levied for water, and an 'excess use' charge has been repeatedly postponed.[29] In Northern Ireland, water and sewerage charges are included in local taxes. England and Wales are slowly moving towards metered consumption, as for gas and electricity supply, with a standing charge (on a per-year or per-day basis) plus an amount per unit of water consumed; 60% of households in England and Wales

From water poverty to water justice

have a metered supply of water but only 13% of all households have a smart meter providing real-time information on usage.[30] Smart meters have only been mandated since 2016, and in most cases a dumb meter is located outside the house under a fixed cover so that households cannot use it to monitor their water use. Some 40% of all households, mainly those living in homes of pre-1990 construction, still pay for water and sewerage according to rateable property value under an arcane system based on 1990s estimates of their houses' rental value. This creates opportunities for householders to game the system, as the moneysaving websites advise that small households and those with fewer members than bedrooms in older houses can gain by shifting to a water meter.[31]

In England and Wales, the average household bill for water and sewerage in the financial year beginning April 2023 was £433 per annum,[32] and as Exhibit 4.1 shows, there is some modest regional variation in the average bill. More to the point,

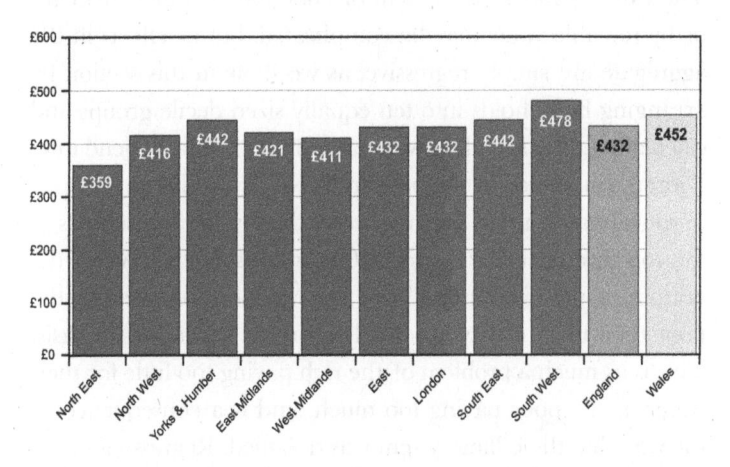

Region	Bill
North East	£359
North West	£416
Yorks & Humber	£442
East Midlands	£421
West Midlands	£411
East	£432
London	£432
South East	£442
South West	£478
England	£432
Wales	£452

Exhibit 4.1 Household annual average water bill by region, 2023[33]

there will be general increases so that the average household water and sewerage bill will increase to £473 per annum in 2024–25[34] (up by some £30 and approaching £10 a week). Further substantial increases are projected in the PR24 2025–30 period, with average bills increasing to around £600 per annum by 2030 for water and wastewater services.[35] National media headlines are understandably all about the average household bill, but that is a statistical construct which conceals as much as it reveals, because the most important point is that water charges are regressive – lower-income households pay relatively more than higher-income households – and this becomes much more of an issue as household bills increase.

4.2 Regressive charging and the limits of social tariffs

The English and Welsh system of charging for water by meter or by rateable value may be complicated, but its effects in the aggregate are simply regressive, as we show in this section by arranging households into ten equally sized decile groups and calculating their spend on water as a share of total expenditure. Poorer households in the bottom three deciles (D1 to D3) pay proportionately more for their water than richer households in the top three deciles (D7 to D10). As households towards the bottom of the distribution have fewer persons on average, the poor typically also pay absolutely more on a per-person basis. This is as much a problem of the rich paying too little for their water as the poor paying too much, and is a consequence of the way that the billing system was designed. Regressivity is an increasingly important social justice issue because, as water bills increase over the next decade or so, the regressive effects will become more vicious (and without solving the industry's revenue constraint problem). Through the water poverty lens, the default fix for charging problems is a national social tariff to address the issue of affordability for the poorest and remove the postcode lottery. But this would also create new problems of take-up by eligible households and cliff-edge inequalities for low- to middle-income groups where entitlement to social tariffs begins and ends. It equally does nothing to make sure that the rich pay more for their water, as they should and could under alternative systems of charging which can equitably raise more revenue.

The regressivity of charging for utilities such as water is not a secret. As long ago as 2013 the National Audit Office published a research report on energy and water bills which showed that these utility charges were regressive – poorer households paid a larger percentage of their income than richer ones – and the NAO added the important point that the degree of regressivity in utility charging was stronger than in general taxation.[36] But this has remained arcane, technical knowledge which has not crossed over into the public domain for several reasons. Politically, reducing poverty has moral appeal and is less adversarial than foundational redistribution, which would take from the rich as well as give to the poor. Technically, it is frustratingly impossible to produce a direct measure which shows how household expenditure varies with household income, because expenditure and income figures come from different survey sources. Official data from one survey does tell us reliably about water spend as a share of total household expenditure, and this is a reasonable approximation of water spend as share of income in the middle of the distribution. But at the top end of the distribution household expenditure is not a good proxy

Murky water

for household income, because higher-income households can and do save, so that they typically earn more than they spend on goods and services. Consequently, for example, households in decile 9 in official expenditure surveys are not necessarily the same households as those in decile 9 in income surveys, and it is not possible to relate decile 9 expenditure on water to decile 9 income.

Our workaround solution is to take spending on water and total expenditure figures for decile groups of households from the Living Costs and Food Survey (LCF), which is a diary-based survey of 4,500 households.[37] The LCF provides robust data on the absolute amount and relative share of spending on water in every decile group. But the caution has to be that calculations from this expenditure source will understate the degree of regressivity. As we have noted, total expenditure is a reasonable proxy for income in middle-income groups but has obvious limits towards the bottom as well as the top of the income range. It is not just that higher-income households do not spend all their income, but that lower-income households may incur debts or draw on inter-household transfers, so that they might spend more than their income. Consequently, any expenditure-based analysis understates the regressive incidence of water charging, especially in the upper deciles. Nonetheless, it is striking that the amount spent on water as a percentage of all expenditure shows consistent regressivity as we move through the deciles from the poorest upwards, both on a household basis and even more so on a per-person basis.

- On a household basis, the water charging system has consistently regressive effects. Data on household spend in England and Wales in 2023, as in Exhibit 4.2, shows that water

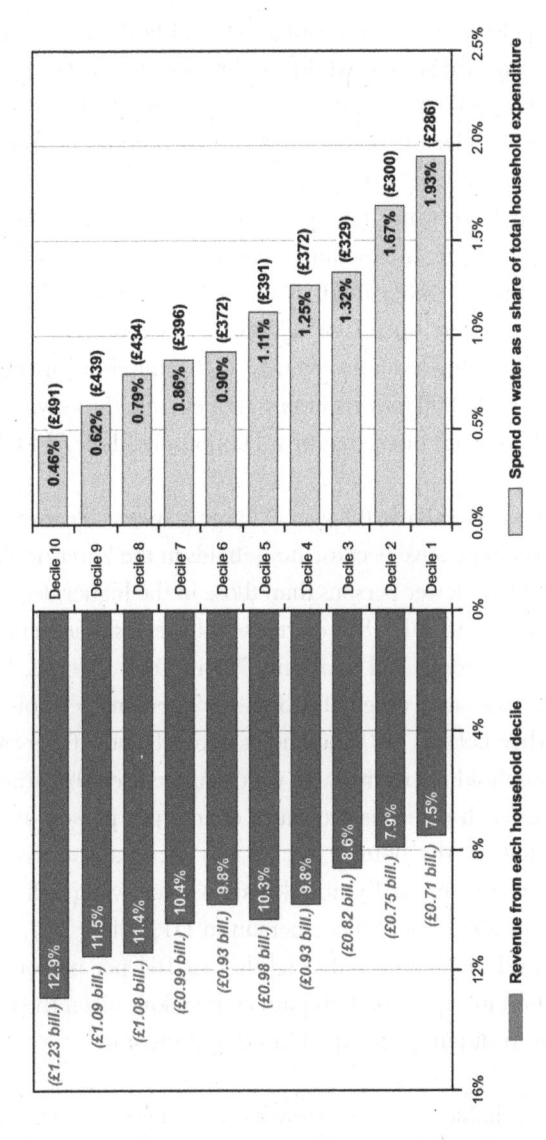

Exhibit 4.2 Household average water bills and the revenue raised in England and Wales in 2023, by decile (total revenue £9.505 billion)[38]

Decile	Revenue from each household decile	Spend on water as a share of total household expenditure
Decile 10	(£1.23 bill.) 12.9%	0.46% (£491)
Decile 9	(£1.09 bill.) 11.5%	0.62% (£439)
Decile 8	(£1.08 bill.) 11.4%	0.79% (£434)
Decile 7	(£0.99 bill.) 10.4%	0.86% (£396)
Decile 6	(£0.93 bill.) 9.8%	0.90% (£372)
Decile 5	(£0.98 bill.) 10.3%	1.11% (£391)
Decile 4	(£0.93 bill.) 9.8%	1.25% (£372)
Decile 3	(£0.82 bill.) 8.6%	1.32% (£329)
Decile 2	(£0.75 bill.) 7.9%	1.67% (£300)
Decile 1	(£0.71 bill.) 7.5%	1.93% (£286)

accounts for a lower percentage of total household spend at each upward step as we move through the deciles from the bottom (D1) households to the top (D10). In D1, water makes up 1.93% of total household expenditure, D2 is not far behind at 1.67% and D3 spends 1.32%. By the time we reach D10, water charges account for only 0.46% of total expenditure. In absolute terms, average D10 household expenditure on water in 2023 was £491, less than double the D1 average spend of £286. If we compare total expenditure, however, an average D10 household spent a total of £108,000 (not including money added to savings), more than seven times that of a D1 household of £14,800 in 2023.

- On a per-person basis, the water charging system is considerably more regressive because households in the lower deciles typically have fewer persons than those in the higher deciles. For example, in D1, 78% of households are single-person households, compared with only 7% in D10. Overall, D1 and D2 have an average of 1.3 and 1.5 persons per household, while D9 and D10 households have 2.8 and 3.0 persons per household respectively. If we calculate the 2023 water spend in each decile as the water charge per person, then individuals in the bottom deciles pay more absolutely as well as relatively. As Exhibit 4.3 shows, the water bill per person in 2023 is £220 per person in D1, falling to £149 in D6 and then rising only slightly to £164 per person in the richest group D10. This pattern is vicious given the top to bottom disparity in expenditure and income.

In an inequitable way, this regressive system could work with little political resistance as long as bills were kept low. The average

From water poverty to water justice

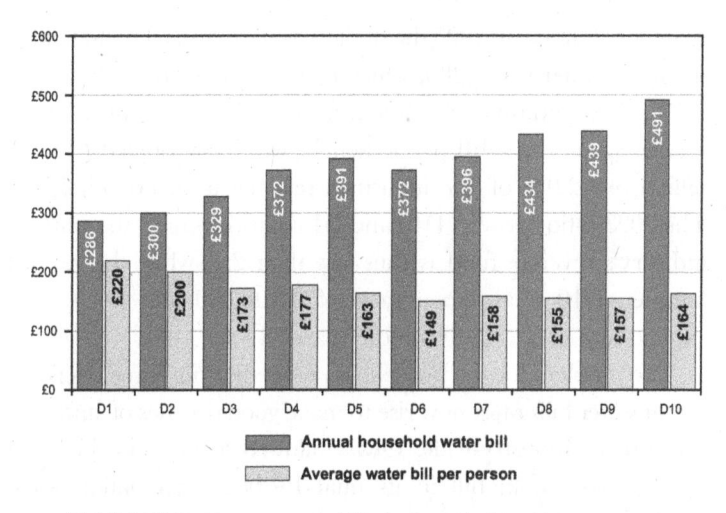

Exhibit 4.3 Average water bill per household and per person, England and Wales, 2023[39]

household bills for 2023 shown in Exhibits 4.2 and 4.3 came at the end of a twenty-year period when bills had not risen in real terms, as we noted in Chapter 2. Consequently, as Exhibit 4.3 shows, the per household spend on water in deciles 1 to 3 is £286–£329 per year, or less than £1 per day. In 2023 poor households were much more exercised by the unaffordability of their utility bills for electricity and gas, which, after the start of the Ukraine war, were much larger than for water. As we have already argued, low average bills have the insidious consequence of creating a revenue constraint and investment rationing. In terms of regressive household charging, this problem of revenue constraint can be understood in a more nuanced way. Revenue constraint is here the result of a failure to charge households in the upper deciles according to their ability to pay and contribute to the revenue fund of the water industry. As

Murky water

Exhibit 4.2 shows, in 2023 the decile 1 average annual household spend on water was £286, which contributed £710 million or 7.5% of the industry's revenue from households; the decile 10 average spend was £491 per household, which contributed £1.23 billion or 12.9% of the industry's revenue from households. The 2023 ratio between D10 and D1 contributions to the water industry's revenue fund is thus less than 2:1, while the ratio between D10 total expenditure of £108,000 and D1 total expenditure of £14,800 is more than 7:1.

The problem for the households in the bottom three deciles is that water bills must now rise to make good decades of underinvestment. Industry-wide, Ofwat allowed a 36% rise in the average household bill in its final PR24 determination for 2025–30, and this enabled planned capital expenditure of up to £104 billion over the five years, which is more than double the £51 billion allowed during 2020–25.[40] This permitted 36% rise is a real-terms increase which, with price inflation, will turn into larger nominal increases when households pay their bills. Thames was offered 35% in Ofwat's final determination, but that was not enough. Like five other water and sewerage companies, Thames Water has appealed against its final determination, though the company indicated that it would like a deferral, hoping that in the meantime it could recapitalise the firm. As we show in Chapter 6.2, major investment projects are now going to be funded by PFI-type deals, with project costs recovered by surcharges on water company bills, so that bills will actually rise by more than Ofwat's headline figures. As a harbinger of surcharges to come, Thames Water customers paid a £26 surcharge on the company's 2024 bill towards the costs of the Thames Tideway super sewer.[41] The investments planned in PR24 for 2025–30 will not adequately deal with the legacy of under-investment and meet the challenge of climate adaptation. Therefore, water bills will have to rise further through the 2030s, especially in the south-east, where a succession of major PFI projects will be required to deal with water shortage and to increase sewage treatment capacity for a population of more than nine million.

It is entirely realistic to suppose that in the later 2030s households in south-east England will be paying water bills double those they paid in 2023; and a doubling of bills is a plausible worst case scenario for all of England and Wales. We do not think that economic growth and rising real incomes will soften this blow. Politicians have no policy levers that could restore 2.5% per annum growth, and they are at the mercy of more unexpected shocks. The bottom three deciles have generally seen negligible wage increases in the period since the 2008 financial crisis, and since 2022 their residual income has been squeezed by energy and food price inflation.

The effect of doubling bills, while retaining the existing regressive system, can then be simulated in a crude but revealing way using the LCF survey expenditure data. Doubling household bills under the existing 2023 charging system means that each decile would contribute the same share of the total paid by all households as in 2023, but the amount of each decile's contribution would be doubled. In 2023 a total of £9.505 billion was raised from household charges for water and sewerage (as shown in Exhibit 4.2); doubling bills would increase the total amount raised to £19.011 billion. To keep the simulation in Exhibit 4.4 simple, we assume that household income and all other expenditure does not change.

With the existing system of charging, a doubling of the total amount paid for water and sewerage by households would produce

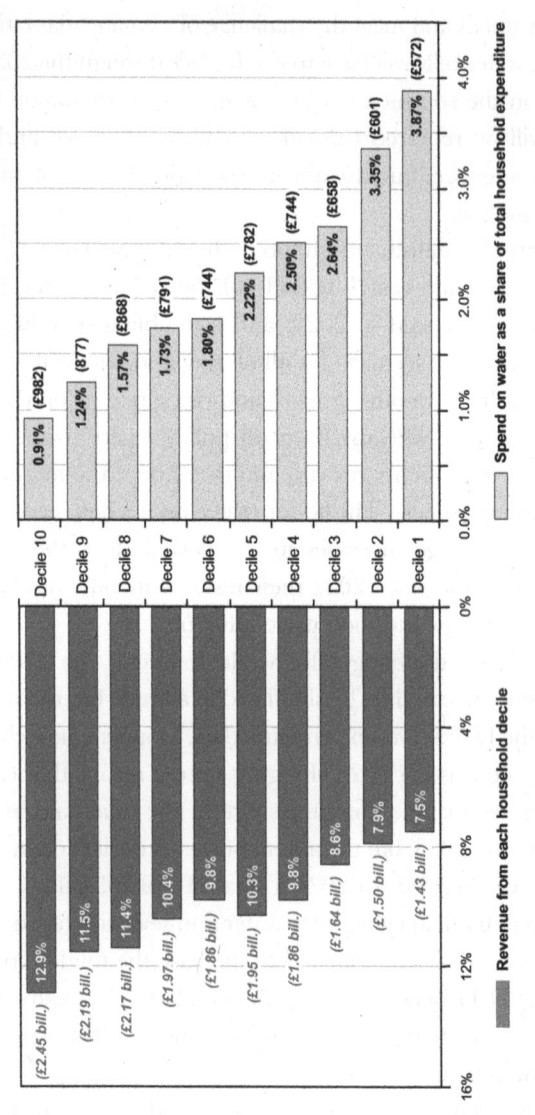

Exhibit 4.4 Simulated doubling of the 2023 revenue raised in England and Wales and the amount paid by each decile (total revenue £19.011 billion)[42]

extremely burdensome bills for poorer households in D1 to D3, while richer households in D7 to D10 would continue to pay a relatively small amount in relation to their total expenditure.[43] In the scenario of doubling the revenue raised, a water bill of £572 is 3.9% of D1 total annual expenditure of £14,800, while a water bill of £982 represents 0.91% of D10 total annual expenditure of £108,000. This simulation is a simple one, but it is good enough to make the important point that the existing system of charging is economically and politically unsuited to the task of the next decade, which is to raise much larger revenues from households. A doubling of bills means that the cost of water would be equivalent to nearly 4% of *all* expenditure by D1, and to more than 2% for all households in the bottom half of the distribution, with some of these spending more than £750 on water per year. In this case, charging would become a political issue, and reform or replacement of the existing scheme is the only sensible option. Before we consider replacement in the next section of this chapter, we consider reform through the introduction of some kind of national social tariff scheme which, as noted earlier, now has widespread support as the fix for affordability problems.

A national social tariff scheme would offer a more unified set of bill reductions for low-income groups who meet defined eligibility criteria. The rebates could be more or less generous, depending on whether the intention is mainly to unify the existing different company-based schemes, or to offer more financial support to a larger group of households. Many West European countries, including Belgium, France, Greece, Italy, Portugal and Spain, have social tariff schemes for energy utility charges. Belgium, for example, offers a social tariff to around 20% of all households, including those on benefits, plus state pensioners, single parents and lower-middle-income groups.[44] So, why not introduce a national social tariff scheme which offers substantial water charge reductions to 20% or more of households? Given the growing interest in such a scheme it is worth highlighting two sets of issues that would both limit its effectiveness and increase the costs for other households, especially those who narrowly miss the eligibility criteria.

The first issue is the well-known take-up problem, which is that not all of those eligible will actually receive the bill reductions. Households in receipt of some welfare benefits could be included automatically in a national social tariff, if systems were effectively joined up, but other eligible households would need to apply. Experience with other schemes such as Pension Credit or Housing Benefit show that up to one-third of those who are eligible do not apply.[45] Water company social tariff schemes are promoted with impressive-sounding 'planned' numbers of customers who will benefit, but there is no assurance that these numbers will be reached.[46] It is already the case that, where the onus is on customers to contact their water company to receive support, many who could be eligible do not apply because of what the CCW terms 'significant barriers'. These can include 'limited awareness' and 'embarrassment', as well as 'perceptions of low impact and limited trust' of the water companies.[47] In principle, barriers of this kind could certainly be reduced by design in a new national social tariff through better information, easier application processes and better customer relations with water companies.

However, the second set of issues are harder to address through design. Social tariffs and price rebates for poorer households address one problem of inequity, that is, a lack of water affordability for the poorest households, but end up creating another problem elsewhere. Most obviously, if they offer a significant reduction to a substantial number of households, they also create an ineligibility cliff edge at the point where the price rebate stops. Those households immediately above this point have to pay the full cost. The impact on those households just above the cut-off for financial support is even greater though, because the revenue lost through the social tariff discounts will be largely covered, not by the companies, but by a bill surcharge on the other customers. In the past, social tariffs have attracted governments and utility companies because they classically cost them very little when they are funded by a transfer between customers. Under the PR24 process, which sets investment and prices for 2025–30, some companies have volunteered a 'shareholder contribution' to help fund the reductions in bills for those customers on social tariffs or WaterSure schemes. Closer examination suggests that very little has changed.

Under PR24 plans there will be large increases in household bills and in the numbers of customers expected to receive rebates. But support from the companies' investors is very uneven: United Utilities and Dŵr Cymru have offered investor contributions of £68.66 million and £63.44 million respectively to partially support social tariffs during 2025–30, while Anglian, Severn Trent, South West, Southern and Wessex have offered no contribution, meaning that the full cost will be borne by customer cross-subsidies.[48] Most of the companies have offered relatively small amounts for 'debt matching schemes' or 'other affordability schemes'. Ofwat presents the total contribution from shareholders to water affordability measures in relation to the return on regulated capital, which is the amount of profit the companies are allowed to generate under PR24. United Utilities and Dŵr Cymru are outliers at 0.48% and 0.42% respectively, while for Wessex, Thames, Anglian and Southern this figure is as little as 0.01%, 0.02%, 0.06% and 0.09% respectively. Under PR24, the allowed return on the regulated equity base is 5.1%,[49] implying that Thames's investors, for example, are contributing less than 1% of their total return, compared with a relatively more generous 8% for United Utilities. The outcome of the plan to offer more customers social tariffs but with very limited contributions from the companies' investors means that the expected average cross-subsidy per customer would be, for example, £55 for Thames, £27 for Severn Trent and £20 for Southern Water, according to Ofwat.[50]

We can see how other customers not receiving any discount are impacted by social tariffs by returning to our scenario of doubling the total amount paid by household customers. For the purposes of illustration in Exhibit 4.5, we assume that the (inevitably) higher water charges are made more acceptable by offering 50% rebates to all households in the bottom three deciles. Deciles 1 to 3 now have their bills halved and the revenue forgone is recovered by a flat surcharge on deciles 4 to 10 (which is not so different from PR24 surcharge plans). The objective of protecting those in the poorest households is met. On this scenario, D1 to D3 expenditure on water is reduced from £572–£658 (as shown in Exhibit 4.4) to £286–£328 with a 50% social tariff rebate. These households would pay no more than they did in 2023, but the water industry revenue from households has still been doubled, as per the scenario. The problem is that this seriously disadvantages the lower-middle-income groups in deciles 4 and 5. Below the cut-off for a social tariff rebate, the D3 spend on water is £328, while above the cut-off the D4 spend on water is £874 and the D5 spend is £912. Both D4 and D5 bills include a charge of £131 to recover revenue lost through the rebate scheme.

From water poverty to water justice

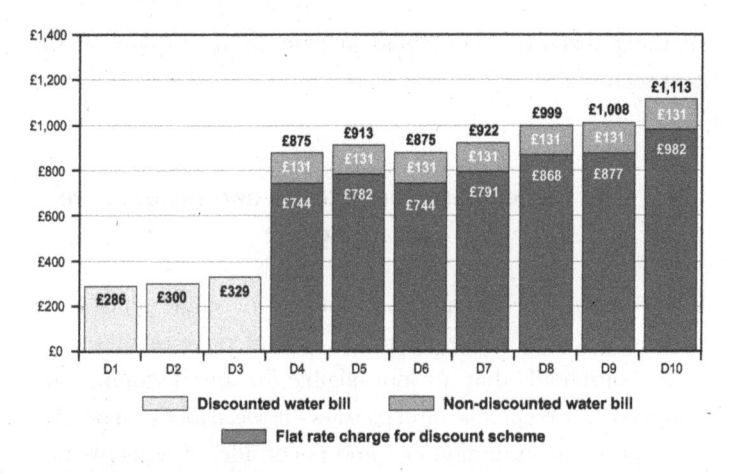

Exhibit 4.5 Simulated household water bills in England and Wales based on doubling the amount raised, with a 50% discounted social tariff for deciles D1 to D3[51]

By introducing a social tariff, policymakers would be simply adding to the multiple take-up and cliff-edge problems which other means-tested benefits have created in the UK. The resulting problem can be managed down by restricting social tariff eligibility to the poorest (such as social security claimants) or reducing the bill rebate, so the poor pay a larger percentage of the bill. Both of these measures undermine the primary aim of substantial assistance to a broad group of low-income households. The other possibility of extending rebates into the middle of the income distribution produces very sharp increases in the bills of the top three deciles but does undermine the scheme rationale of providing targeted assistance for low-income households. In which case, if social tariff has so many unintended and inescapable negative consequences, it would be better to institute a more radical reform and consider the case for flat-rate or progressive charging related to household income, as we explore in the next section.

4.3 The justice case for flat-rate and progressive charging

If and when the UK adopts a national or single social tariff scheme, it will most likely load the costs of price rebates on to those households that are not eligible for any discount. This suits the water companies and placates the social lobby on poverty, but represents an undisclosed and ill-considered redistribution by stealth through transfer between customer groups. This kind of scheme is a poor substitute for a progressive charging system which in an explicit and deliberate way delivers social justice by reducing water costs as a share of expenditure in deciles 1 to 3 *and* by increasing water costs as a share of expenditure in deciles 7 to 10. Varying the charge that households pay according to household income is the most straightforward way of securing the justice result. This section uses new simulations of flat-rate and progressive systems of charging which rework the 2023 expenditure data presented earlier in the chapter to explore counterfactual alternatives and model the outcomes. These simulations suggest that alternative ways of charging would not only deliver more equitable outcomes, but also would allow the water industry to raise more revenue and address its business model problems. With this point made, we go on to consider the technical difficulty, social relevance and political feasibility of household income-related charging where the balance between household winners and losers – and the reactions of those who lose – become important.

From water poverty to water justice

The first step is to show that a simple flat-rate charge, with all deciles of households paying the same percentage share of income, produces a much more equitable result when tasked with raising the actual water industry revenue of 2023. A flat-rate charge on households can also raise double that revenue or more, with acceptably equitable results. To be clear, with a flat-rate charge, households do not all pay the same fixed amount, as in a poll tax or the BBC licence fee. Instead, all households pay the same flat rate levied as a percentage of household income. This means that high-income households pay the same relatively but much more absolutely than low-income households, because the fixed rate is applied to a much larger income base. In our counterfactual simulations we have to work from household expenditure (not income) by decile, for the reasons explained in the last section. In our new simulations we start with a fixed rate of 0.86% of total expenditure, which is the average percentage of total expenditure that households in every decile would have to pay to raise the same £9.505 billion of revenue obtained from all households in 2023 (Exhibit 4.6). We then apply 1.72% as the average percentage of their total expenditure that all households would have to pay to raise double that revenue of £19.011 billion (Exhibit 4.7).

The payments made by households in each decile in these new simulations are directly comparable with those in Exhibits 4.2 and 4.4, which summarised first how the existing water charging system actually raised the 2023 revenue and then how that existing charging system might raise double the amount from households. The first new simulation in Exhibit 4.6 models the counterfactual case of raising actual 2023 revenue using a flat rate instead of the existing regressive system: Exhibit 4.6 shows that the average bill for the poorest 10% of households

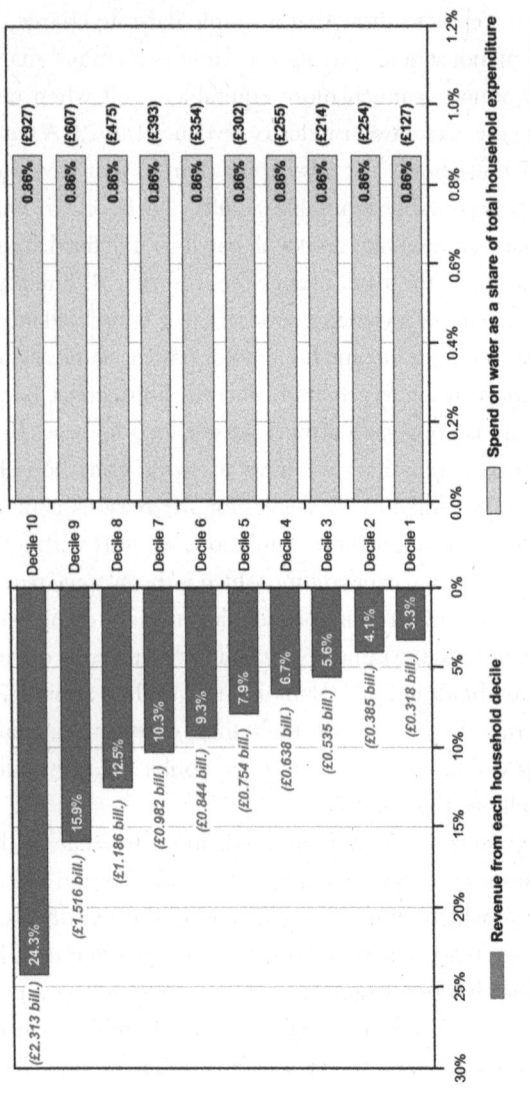

Exhibit 4.6 Simulated 2023 water revenue of £9.5 billion, raised using a flat-rate charge of 0.86% of household total expenditure, by decile[52]

■ Revenue from each household decile
□ Spend on water as a share of total household expenditure

Decile	Revenue	Share	Spend on water
Decile 10	(£2.313 bill.)	24.3%	0.86% (£927)
Decile 9	(£1.516 bill.)	15.9%	0.86% (£607)
Decile 8	(£1.186 bill.)	12.5%	0.86% (£475)
Decile 7	(£0.982 bill.)	10.3%	0.86% (£393)
Decile 6	(£0.844 bill.)	9.3%	0.86% (£354)
Decile 5	(£0.754 bill.)	7.9%	0.86% (£302)
Decile 4	(£0.638 bill.)	6.7%	0.86% (£255)
Decile 3	(£0.535 bill.)	5.6%	0.86% (£214)
Decile 2	(£0.385 bill.)	4.1%	0.86% (£254)
Decile 1	(£0.318 bill.)	3.3%	0.86% (£127)

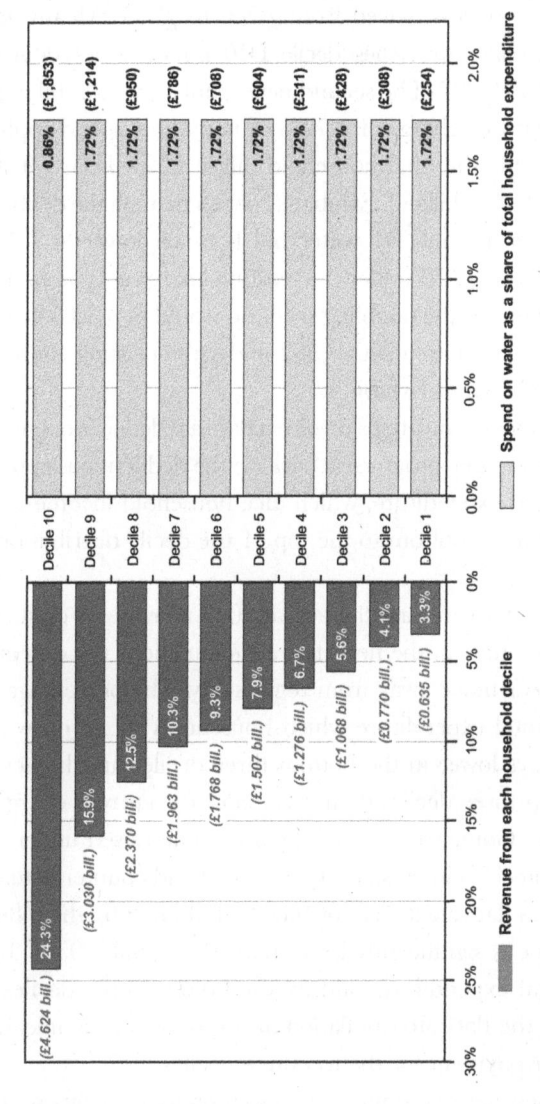

Exhibit 4.7 Simulated doubling of the 2023 water revenue to £19 billion, raised using a flat-rate charge of 1.72% of household total expenditure, by decile[53]

(D1) is more or less halved from £286 to £127. Meanwhile, the average bill for the richest decile, D10, is more or less doubled from £491 to £927. The second new simulation in Exhibit 4.7 models a flat-rate charge of 1.72% of household spend applied to raise double the total household bill in 2023 (from £9.505 billion to £19.011 billion). Exhibit 4.7 shows that, on the doubling scenario, the average D1 water bill is much lower at £254, compared with £572 under the existing charging system. The D10 water bill on the doubling scenario would be £1,853 using the flat rate, compared with £982 under the existing charging system (as shown in Exhibit 4.3).

The two new simulations involve a flat-rate deduction, but the outcome is broadly progressive because the deduction is applied to household expenditure, which (like household income) rises steeply from the bottom to the top of the decile distribution.

- Whether it is raising the actual 2023 revenue or doubling that revenue, in the new flat-rate simulations the outcome is a clear move away from regressivity. The percentage of their total expenditure which households spend on water is always lower in the bottom three deciles and higher in the top three deciles than it is under the current charging system. Comparing actual 2023 revenue raised under the existing charging system (Exhibit 4.2) and counterfactually under a flat-rate system of 0.86% (Exhibit 4.6), the 0.86% flat rate is significantly lower than the actual 1.93–1.32% of total expenditure paid by the bottom three deciles in 2023; the flat-rate simulation also produces a significantly higher payment for the top three deciles.
- The new flat-rate simulations also produce a nicely progressive result in terms of the amount that each household

decile contributes to the total raised. Raising the actual 2023 revenue on a 0.86% flat-rate basis means that total revenue contribution by D1 households is more than halved from £0.7 to £0.3 billion, while D10 households make a more significant overall contribution to total revenue of £2.3 billion (up from £1.2 billion with the existing charging). If we apply a 1.72% flat rate to raise double the 2023 revenue, D10 households make a major revenue contribution of £4.6 billion, while D1 households contribute only £0.6 billion of revenue. In both new simulations, the burden is shifted on to the broadest shoulders. Under the present charging system D10 contributes just 13% of total revenue, but the flat-rate charge increases this to 24% of revenue.

- Given the inevitability of substantial bill increases, the last and most important point is that the new simulations show that flat-rate charging can raise substantially more revenue without overburdening the bottom three deciles. As we saw in the last section, this is exactly what the existing system of charging cannot do. In our flat-rate simulation of raising double the actual 2023 revenue, deciles 1 to 3, like all other deciles, pay 1.72% of their total expenditure; but under the current charging system they would pay 3.9% to 2.6% of expenditure.

In a second step we can move on from flat-rate simulation to a progressive rate under which poorer households pay a lower share of expenditure and richer households pay a higher share of expenditure. A progressive rate can, of course, be varied across the deciles to produce different results according to whether the taper is gentle or steep. This opens up the possibility of fine tuning according to political choices about how much relief to offer households at the bottom of the income distribution by loading charges on to households at the top of the income distribution.

In our new simulation of a progressive rate of charging we illustrate the potential with a worked example, where the taper is set so that water bills can be reduced to negligible levels for the bottom decile and for most of those in the bottom three deciles. In this new simulation we meet the objective of reducing the annual bill of the poorest households to a point – £38 in D1 and £50 in D2 – where it is hardly worth the administrative expense of collecting the revenue. This result comes with the heavy lifting done by the top three deciles all paying bills which are significantly larger. But if we consider ability to pay, the top three deciles in this new simulation do not face exorbitant and unreasonable increases from the current actual bill in 2023. The D8 spend on water rises from £434 to £533 on annual household spend of £55,000, the D9 spend on water rises from £439 to £857 out of an annual household spend of £71,000, and the D10 spend on water rises from £491 to £1,417 on an annual spend of £108,000. There is something of a cliff edge between deciles 9 and 10, with water spend in D10 around £400 higher than in D9, but this reflects income-related ability to pay.[54]

Our flat-rate and progressive simulations show that there are strong social equity and business model arguments for shifting to charging on a basis related to household income. Charging according to household income has a clear logic in a society of great income inequalities. Water bill increases which are modest in relation to large incomes in the top three deciles would cover very dramatic reductions in the water bills of poorer households, both in absolute and relative terms. And if the introduction of progressive charging was delayed by inevitable wrangling about how progressive the system should be, a simple flat-rate system would produce large social and economic benefits that are unattainable under the present regressive system of charging. An explicitly progressive basis for raising the revenue required to maintain and upgrade the infrastructure can also deliver household justice and recognise the special character of water as a public service whose provision of clean and wastewater services shapes ecological outcomes and is increasingly part of climate resilience.

If income-related charging is attractive, it would not be a small change. Charging for water according to household income is technically feasible but would almost certainly need to be part of a comprehensive reform of utility charging. The practical obstacle is that the UK government would need information on all household incomes before it could vary water and other utility charges on this basis. The UK government already has data on the income of many households scattered across different databases. For example, 6.4 million people (including 2.6 million in employment) in early 2024 were on Universal Credit and all of these will have provided household income information.[55] Almost 8 million households claim Child Benefit, where couples in a relationship have to declare the income of both partners,[56] and, in England, 1.5 million applied in 2023/24 for a student loan, which is granted after a declaration of parental income.[57] All these state aids depend on household form filling, which many more already do on an individual basis: 12.1 million (of the 42 million adults over 15 in the UK) are now expected to file a self-assessment tax return[58] as the basis for paying individual income tax. If the UK government wanted to, it could justify acquiring annual returns on household income on the grounds that this could be used for progressive innovations in water charging and in other areas. Charging according to household income would be a major change, but the technical problems are soluble.

This is good news because the social justice case for radical innovation in charging for utilities and in other areas is becoming more pressing. Shifting the basis of utility charging on to *household* income would recognise two new socio-economic realities: first, secular changes which have led to the demise of the household with a single (mostly male) 'breadwinner', and the emergence of both low- and high-income dual-earner families; second, the conjunctural pressures of the cost-of-living crisis with much higher prices for the on-market foundational essentials of energy, food, housing and transport.

- As female workforce participation has increased, many low- and high-income households have become units for consolidating the incomes of two or more wage earners. The upper deciles are largely populated not by elite figures but by ordinary two-income managerial and professional couples. The proliferation of low-wage and part-time service jobs has also put a second (usually female) minimum wage worker into many hardscrabble households. But the UK system of taxing income, from the nineteenth century and Beveridgean social insurance in the 1940s, presupposed a male head of household as sole earner; indeed, for Beveridge, a wife was covered by her husband's contributions. The introduction of Pay as You Earn (PAYE) in 1944 extended the income tax net to take in working-class male heads of household who could not settle an annual tax bill in arrears, as their middle-class counterparts could. This issue remains relevant for income tax on low earners, and undoing PAYE would be a step too far. But more broadly, the single-earner focus has become obsolete given increased female participation in paid work. The relevance of the household is already recognised in state benefit calculations, and this could now be extended to utility charging.

- The cost-of-living crisis, with higher energy, food, transport and housing costs, has (as we noted in our last book) squeezed household spending and highlighted problems caused by existing systems of utility charging. Utility bills for electricity and gas (and increasingly for water) are charged on a household basis, classically with the household paying for consumption on a metered connection, and with a standing charge. The logic of metered connection is regressive charging in relation to household income, and is even more regressive if we consider the charge per person, as with water. Gas and electricity bills have already risen significantly since 2022 and are unlikely to revert to pre-Ukraine war levels, so household budgets are under pressure even before water charges increase from 2025. Without information on household incomes, the UK government's policy response to the energy price hike was an untargeted and effectively regressive price cap which set a maximum electricity and gas price per kWh for all households regardless of income.[59] Just as we have increased numbers of households in water poverty, so there are also growing numbers in energy poverty[60] and food insecurity.[61] These different kinds of 'poverty' are often looked at separately, though of course they are substantially overlapping and interrelated, and reflect the underlying problem of low incomes. What is needed is an integrated approach to utilities, and household income charging is the policy lever that can deliver that.

Murky water

If charging according to household income is more equitable, is technically practical, and fits current socio-economic realities, the serious obstacles are all political ones. This is especially so in the case of water, because the system of charging for water is entirely a matter of domestic political choice at the UK level in Westminster and Whitehall. In the case of electricity or gas, market prices (before policy intervention) are directly set by, or indirectly related to, the international commodity prices of oil and gas. So the policy question in electricity and gas is whether and how domestic policy should mediate the effects of uncontrollable external events. In the case of water, the costs of system maintenance and renewal are entirely domestic and the allocation of those costs to households in different income groups is a matter of policy choice by government, which could change the basis for household charging, or indeed abolish direct household billing for water and recover the costs through national or local taxation as in the Republic of Ireland or Northern Ireland. The obstacle to change is Westminster politicians, who are traditionally reluctant to propose radical changes in the tax system because they fear blowback from those who would lose from such changes. So, the question must be: is blowback from the financial losers of this change a decisive obstacle in the case of water?

From a narrow political point of view, change in the charging system is less about abstract considerations of justice and more about the practical balance between the number of households that win by paying smaller bills and those that lose by paying higher bills. This issue is clarified for the water industry in Exhibit 4.8. This provides a decile breakdown which compares the bill paid and the percentage of total expenditure spent on water when the 2023 revenue of £9.505 billion is raised under three

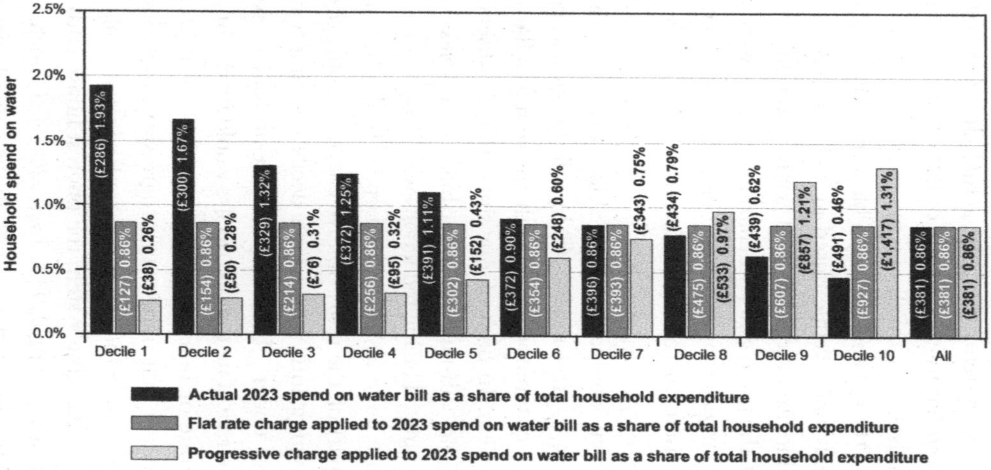

Exhibit 4.8 Analysis of household expenditure on water, contrasting the current system in 2023 with flat-rate and progressive simulations[62]

different charging systems: first, the actual charging system in 2023; second, our flat-rate simulation of 0.86% of total expenditure for all households; third, our progressive-rate simulation which reduces D1 to D3 bills to negligible levels. The comparison applies to the actual revenue raised in 2023 in each case, but the bill relativities and the position of winners and losers would be the same if the three different charging systems were tasked with raising a larger amount of revenue from households, as is now the case where bills are increasing to accelerate investment in physical assets.

The three-way comparison in Exhibit 4.8 shows that the balance between winners and losers is broadly favourable, because winners greatly outnumber losers with any shift to a flat rate or progressive system. In both of our simulations, households in deciles 1 to 5 pay substantially lower bills and deciles 6 and 7 pay much the same. The bill reductions with flat-rate or progressive charges are large at the bottom end. Deciles 8, 9 and 10 pay more as part of the redistributive effect, but only deciles 9 and 10 pay significantly more, with decile 8 facing a modest increase of just under £80 on the flat rate and £137 extra on the progressive rate. The top two deciles would no doubt protest noisily if flat-rate or regressive charging were introduced. But it should be noted that under our progressive simulation D10 pays 1.31% of a large total expenditure for water, while under the actual 2023 charging system the poorest households in D1 to D3 are already paying 1.93% to 1.32% of a much smaller total spend. The choice of rate and, in the case of progressive charging, the taper can of course be modelled and adjusted to avoid cliff edges and perverse outcomes.

The conclusion has to be that the balance between winners and losers should not be a political obstacle to reform of water charging for social justice. Charging according to household income is a viable alternative and should be within the realm of the politically possible. But politics is somehow or other stuck in Westminster and Whitehall, and alternative methods of charging are not on the policy agenda. This is part of a broader problem. Most obviously, the Labour government elected in 2024 is reluctant to contemplate special administration, let alone nationalisation of water companies, even in the case of Thames Water which, at the time of writing, was trying to distress borrow an additional £3 billion at a hefty 9.75% interest rate plus fees, all of which will have to be paid for by customers.[63] The next chapter's analysis of issue management and regulation looks at how and why the Westminster and Whitehall politics of water is stuck such that it is not about solving problems.

Chapter 5
Failure of political and regulatory control

Introduction

This chapter analyses a threefold failure of political and regulatory control: Defra as the responsible government department has diverted into issue management of symptoms; Ofwat as 'economic regulator' has not effectively questioned business models and controlled financial engineering; and the Environment Agency has been ineffective but has provided political cover for Defra ministers.

Section 5.1 explores a failure of political control, where Defra can be blamed as the department responsible for the water industry. Defra's failure is unsurprising when so much of UK electoral politics works as an echo chamber, responding to public opinion in a superficial and narrow way without engaging with underlying problems. Westminster politicians respond to opinion polls and focus group priorities, which divert them into issue management. Policy initiatives then signal the desire to improve but cannot

Failure of political and regulatory control

deliver improvement when issue management leads politicians to focus on symptoms, not underlying problems. In party manifestos before the 2024 General Election and in Labour government initiatives since the election we find echo politics about sewage in rivers, corporate misbehaviour and fat cat salaries. Disappointment was inevitable when, before the election, official reports had made the case for more radical action and, after the election, civil society groups expected more than issue management.

Section 5.2 analyses the dismal record of Ofwat as the independent regulatory agency directly responsible for two decades of physical under-investment in water, during which it allowed financialised investors to load companies with debt. The mitigating circumstance is that Ofwat's generic brief as 'economic regulator' of a privatised monopoly was to balance low prices for consumers with adequate returns for regional water companies. This brief did not take into account the activity characteristics of the water industry outlined in Chapter 2. It also left Ofwat with a business trilemma where it had to choose two out of three objectives: low bills, adequate physical investment and acceptable financial returns. Ofwat negligently sacrificed physical investment and at the same time did nothing to curb financialised investors such as Macquarie which were extracting double-digit returns for shareholders by loading companies with debt. The fundamental problem was and is that Ofwat has a level of economic expertise which is not equal to the accounting task of analysing business models and controlling financially engineered businesses.

> *Section 5.3* explains the limitations of action by the Environment Agency. It has never been adequately resourced for its multifarious regulatory duties inside and outside the water industry, and the Agency is 'sponsored' by Defra, which sets the environmental policy framework within which it works. In the 2010s the Agency acquiesced in its own ineffectuality by accepting the operator self-monitoring system, which allowed water companies to under-report pollution incidents, and the Agency is only now recovering from its irrelevance by slowly putting more resources into inspection and monitoring. Some of what the Agency does promises more than it delivers. The long overdue fitting of event duration monitors to storm overflows is a step forward. But at the time of writing in early 2025, spills at a specific site cannot be related to local rainfall so as to identify egregious 'dry spills'. In political terms, however, the Environment Agency has been useful because it has provided cover for successive Defra ministers facing criticism or questions on environmental issues.

5.1 Political control as issue management

If water is in a mess, this is the result of a failure of political control. Successive ministers and Defra are immediately to blame because they are formally responsible for the water industry. But their failure is symptomatic of a broader inability of the mainstream parties and the political class in Westminster and Whitehall to focus on problem definitions and solutions. In Chapter 6 we will argue that this broader failure is part of a

Failure of political and regulatory control

much larger set of UK problems beyond water, where it is difficult to get things done across many areas of public life because of elite power under the tutelage of finance. However, the failure of control in water more immediately relates to the way that much of UK electoral politics works as an echo chamber, responding to public opinion in a superficial and narrow way, without engaging with underlying problems. Senior politicians in all the main parties operate in response mode as they construct an electoral offer on the issue in question, responding to intimations about what the public thinks and why they think it. Issue management is, then, about political elites signalling the desire to improve, not delivering meaningful improvement. All this is obvious in water in recent years, because the behaviour of companies and the crisis of water quality were public concerns in the run-up to the 2024 General Election and afterwards for the newly elected Labour government.

Echo chamber electoral politics is not new, but it has been powerfully boosted by the way in which UK politics was reformatted after 1990, through the use of opinion polls and focus groups for sampling and reading public opinion. Sample-based opinion polling had figured strategically in Westminster politics at least since the 1960 publication of Mark Abrams et al.'s *Must Labour Lose?*[1] But discussion-based focus groups only crossed over from market research into political communication in the 1990s with New Labour's aim of repositioning the party to attract voters with new social and economic attitudes and group identities.[2] Subsequent leaders of major Westminster parties have all relied heavily on private polling and focus groups. This creates a new role for experts, who typically work by relating poll and focus group findings to sociographic sub-group identities that reflect values and beliefs as much as economic position. This was what

Murky water

Philip Gould did for Tony Blair and what Deborah Mattinson and Clare Ainsley have done for Keir Starmer.

Frontline Westminster politicians have been understandably reticent and defensive about how this back-office advice influences their front-office retail political offers. The technicians have argued that, as Gould put it, their techniques are 'part of a necessary dialogue between politicians and people'.[3] Such techniques do not have an invariant set of effects; clearly, it all depends on who uses the techniques and how. However, they have a powerful and generally limiting effect on centrist politicians who are generally happiest going with the flow of mainstream opinion and upsetting as few electors as possible. The stream of opinion polls and focus group findings bias elected politicians towards response mode. Politicians remain formally in authority, but in effect they accept a subordinate or follower role, so that opinion poll findings and focus group framings drive their actions within broad limits set by their party's political identity. Our argument is that this form of issue management has now become the default mode in Westminster politics, so that it shapes the response to the water industry crises, even when water does not figure as a top ten, let alone a top three issue of public concern.

Philip Gould claimed that the new techniques were powerful because they allowed elected politicians to tap 'a rich vein of empirical common sense'.[4] However, there is a problem here because, in echo chamber politics, Westminster leaders are catching up with what the general public think, and look to address the media-fed headline issue, not the underlying problems. The result in water is a classic case where the dynamics of echo chamber politics are played out so that electoral politics becomes issue management.

Failure of political and regulatory control

1 In terms of the problem definition, scandal generally figures prominently, with public indignation a powerful driver. In the case of water, public disgust and civil society agitation focus on raw sewage in rivers and coastal water, corporate misbehaviour, and 'fat cat' salaries, while challenges such as regressive charging or climate adaptation are much less visible. Through the political echo effect, sanitation is therefore the issue to be managed, with water management very much in the background, which means that the underlying causes and preconditions for solving the narrowly defined problem are not identified.

2 The task of echo chamber politicians is to signal that they 'get it' and then to manage public indignation about an issue by constructing a narrative of purposive action. Political action is not about tackling a soluble problem with root causes but about accepting a set of priorities that are turned into visible actions, suggesting that problems are being tackled in a purposive way. While there is tough talk, there is also conciliation of other interests, with lobbyists pressing their case. In the case of water this concern with the appearance of purpose can be tracked in the party manifestos for the 2024 General Election and in the subsequent policy initiatives of Steve Reed as Secretary of State at Defra in the new government.

3 The limits on political action in long-established political parties are set by the identity of the party managers in economic and social terms. New right parties such as Reform can partly escape the identity constraint because they can fuse traditional left and right economic and social positions to electoral advantage. Others are more fixed. In the case of water, although 80% of the public want public ownership

of water, this option is ruled out by the government, whose identity is muddled, but whose positioning depends on rejecting the old-fashioned leftism of the previous party management under Jeremy Corbyn.

4   Issue management is compellingly attractive for politicians without values or analysis, but issue management is often self-defeating. Ostentatiously dealing with symptoms generally produces disappointment about improvement not being delivered and problems not being addressed and resolved. In the case of water, disappointment was inevitable during and after the 2024 election. Official bodies and select committees had warned about inadequate government policies long before the election campaign, and engaged civil society groups were soon afterwards disappointed with the policy initiatives of the new government.

All these points can now be developed using the party manifesto positions on water for the UK's 2024 General Election and Defra's policy initiatives in the newly elected government. But the first point to make is that Westminster politicians were, from 2022 onwards, warned about the gravity of the problems in water in a slew of reports from parliamentary select committees and two relatively new statutory bodies, the Office for Environmental Protection (OEP) and the National Infrastructure Commission (NIC). The message of these reports was that the government's own targets and ambitions were not being met, and water and sewerage infrastructure was not being prepared for future needs and challenges.

- The House of Commons Environmental Audit Committee inquiry in 2021 led to the publication of a 2022 report, *Water Quality in Rivers*, and a progress report in May 2024.

Failure of political and regulatory control

This report set out the scale of action required to address the 'mess' in stark terms: 'A step change in regulatory action, water company investment, and cross-catchment collaboration with farmers and drainage authorities is urgently required to restore rivers to good ecological health, protect biodiversity and adapt to a changing climate.'[5]

- In 2023 a House of Lords Industry and Regulators Committee inquiry into the sector diagnosed clear government and regulatory failure: 'The Government and regulators including Ofwat and the Environment Agency have not approached the key issues facing the sector in a joined-up way, including reducing water pollution and securing future supply.'[6]
- The NIC was the body charged with giving 'expert, impartial advice' to government on 'future infrastructure needs and solutions'.[7] In its second National Infrastructure Assessment of 2023, the Commission concluded that 'current government action on resilience is insufficient'.[8] Specifically, it noted that delayed publication of resilience standards meant that these would not drive investment in asset resilience in the water sector until the 2034 price review.
- The OEP has been, since Brexit, the 'public body that protects and improves the environment by holding government and other public authorities to account'.[9] In January 2024 the OEP surveyed the nine assessable goal areas of the government's Environmental Improvement Plan and concluded that the government was 'largely off target' in seven areas and 'partially off target' in the remaining two.[10] In all areas 'progress and prospects are impeded by the lack of an effective and transparent delivery plan'.[11] The accompanying press release starts by noting that 'deeply concerning

failures to properly implement regulations designed to protect rivers, lakes and coastal waters in England mean key targets for improvement will be missed'.[12]

The straightforward implication of these – and many other – reports was that radical policy action was required. Yet the power of echo chamber electoral politics is such that the mainstream parties defaulted into issue management, with outrage about storm overflows centre stage, and all parties glossed over the difficulty of making reform work.

It was the non-centrist outlier parties – the Greens and Reform – who majored on public ownership in their 2024 manifestos, as Labour under Corbyn had done at the previous 2019 election. In both cases, change of ownership was presented as the solution, without recognising the difficulty of making it work. The Greens simply promised in one sentence of 20 words to 'end the scandal of sewage pouring into our rivers and seas by taking the water companies back into public ownership'.[13] Reform proposed a new model in all the utilities (including water) with 50% 'public ownership' and 50% ownership by UK pension funds, to 'rebuild crumbling infrastructure', without giving any further detail.[14] Labour in the run-up to the 2019 election did at least have a much more developed proposal for how the water companies would be bought out and how a 'publicly owned water system' would be governed.

The Liberal Democrats pledged to transform water companies into 'public benefit companies', with local environmental groups given a 'place' on the board, which would create for-profit private companies that also delivered broader social benefits of their choosing. The Liberal Democrat manifesto also proposed replacing Ofwat with 'a tough new regulator with powers to

Failure of political and regulatory control

prevent sewage dumps'.[15] If the two traditional major parties – Conservative and Labour – similarly did not go near public ownership, they were more timid in that they both postponed the issue of reorganising the regulation of the private industry. The incumbent Conservative government postponed regulatory reform until after 2030, when the Ofwat price review criteria would be changed to encourage catchment-area and nature-based solutions. In opposition, Labour in 2023 had internally discussed reorganising water regulation to create one unified agency.[16] However, in their 2024 manifesto they pledged only to put 'failing water companies under special measures to clean up our water'.[17] After winning the election, the new government postponed any major decision by setting up the Cunliffe Commission, whose recommendations it could choose whether or not to implement.

The power of the echo chamber is demonstrated by the neglect of charging and revenue constraint issues, and by the way in which the problem of storm overflows is foregrounded and treated. The Liberal Democrats deserve some credit as the only party to propose specific social and environmental measures, including a single social tariff and sustainable drainage schemes.[18] But equally they did not recognise that the industry's profits were certainly not large enough to fund investment in clean-up even before they had been skimmed by the manifesto proposal for a 'sewage tax' on profits. This assumption that the industry's business model had adequate profitability and no revenue constraint is (and has been) the default across the political spectrum. When it proposed nationalisation in 2019, the Labour Party did not confront the problems of funding replacement and enhancement investment. It was simply assumed – without any justification – that water industry operations would break even or generate a surplus which would be retained.[19]

Murky water

The clear sense from the manifestos of the three major parties was of parties in response mode to storm overflows and raw sewage in rivers and coastal waters. Just as significant was the way in which they addressed this problem, with seeming agreement that the problem is one of corrigible misbehaviour by water companies, and not also a structural problem about revenue and the business model. The appropriate remedy for misbehaviour was a series of new conduct requirements and extra punishments for delinquent executives and companies.

- As the longstanding party of government, the Conservatives recognised that listing their achievements would not appease indignant electors. They promised extra punishments as they would 'hold companies to account' with unlimited fines, while banning executive bonuses where delinquent companies committed a serious criminal offence.[20]
- The Liberal Democrats heaped blame on the Conservatives for 'letting water companies get away with pumping filthy sewage into our rivers and onto our beaches', and proposed stricter enforcement, more monitoring of water quality and new 'legally binding targets'.[21]
- Labour, as the party positioned to win the 2024 election, addressed water in a pre-election document titled 'How Labour will tackle sewage spills'. This proposed much the same extra punishments as the Conservatives. Labour proposed to 'block the payment of bonuses to executives who pollute our waterways', with 'criminal charges against persistent law breakers' and for 'water bosses who oversee repeated law breaking'; plus, more effective monitoring with 'severe and automatic fines for wrongdoing'.[22]

Failure of political and regulatory control

After Labour won the election, it became clear that their echo chamber pre-election pamphlet was not cover for a thought-through radical plan for post-election action. What the new Defra Secretary of State Steve Reed did was to turn up the rhetoric: 'We'll charge water bosses and ban their bonuses until they fix this toxic mess', screamed the headline over an article he wrote for the *Daily Mail*.[23] The result was the Water Special Measures Bill, which enacted the punishment measures proposed by Labour before the election and added two customer-friendly proposals.[24] New 'customer panels' in water companies would have 'powers to summon board members for questioning to make sure they are meeting customer obligations'. Cash compensation for households and businesses was to be doubled 'when basic water services are affected', such as when a 'boil water notice' is issued.[25]

Defra's announced 'crackdown' in the Water Bill was carefully designed so that it appeared tough, but it contained no provisions that the water companies could not easily live with. Water company executives could be 'sent to jail', but only in the very unlikely event that 'their water companies fail to cooperate or obstruct ... investigations'.[26] The Bill banned the payment of performance-related pay, including bonuses to CEOs and senior executives, 'unless they meet high standards', though it imposed no restriction on high rates of basic pay. Ahead of the Water Bill published soon after the 2024 election, it had been announced that funds allocated for investment would be ring-fenced and returned to customers if they were not spent on upgrades, so that underspend could not be distributed as interest or dividends.[27] But there were no plans to intervene on the payment of dividends, and interest charges will continue to rise as the industry's debt increases, as envisaged in companies' PR24 plans and Ofwat's final determination.

The 'crackdown' reflected the new government's approach to the industry, which set limits on what could and could not be changed. For Steve Reed, the new government's approach involved two desiderata: that 'customers and the environment always come first' and that 'the water sector can attract the investment that's needed'.[28] In the jargon of the financial sector, the industry had to remain 'investable', meaning that the industry must offer competitive returns to existing and new investors. On this basis, restructuring a company such as Thames Water should, if possible, be avoided because that might discourage private investors in the industry as a whole. If restructuring becomes necessary, it would (under existing special administration regulations) operate like private-sector insolvency to allow the exit of a private-sector 'going concern'.[29] All this fits with Reed's claim that 'the [industry's] failure isn't linked to ownership, it's linked to regulation and governance'.[30] The outstanding issues regarding regulatory reorganisation and the attraction of new private project investment were handed over to the Cunliffe Commission in late 2024, with a brief to consider everything except re-nationalisation.[31] Significantly, the Commission chair Jon Cunliffe was a safe choice as a retired Treasury official and Bank of England deputy governor, not an engineer or scientist.

In the period after the 2024 General Election, the limits of issue management were quickly exposed by civil society groups which had hoped for more from a new Labour government. When the provisions of the Water Bill were being discussed in autumn 2024, leading campaigner Feargal Sharkey said, 'I see nothing so far that is actually showing anything resembling a strategy and a cohesive plan to deal with either the sewage crisis,

Failure of political and regulatory control

the environmental crisis, the agricultural farmyard pollution or indeed the over abstraction of chalk streams.'[32] After the Bill had been published, one campaigner told the BBC that it was 'window dressing'. Charles Watson, chair and founder of River Action, welcomed the government's acknowledgement of the 'scale of the problem', but added that 'If the Secretary of State believes that the few one-off actions announced today, such as curtailing bosses' bonuses, however appealing they may sound, are going to fix the underlying causes of our poisoned waterways, then he needs to think again.'[33]

5.2 Ofwat: economic regulation of business

It suits centrist politicians to blame regulation for industry failure because that distracts from the issue of ownership. Nonetheless, the two main regulators – Ofwat and the Environment Agency – must in different ways take their share of the blame. Ofwat as 'economic regulator' is directly responsible for years of under-investment and the failure to control financialised owners. In Ofwat's defence we could plead diminished responsibility or at least mitigating circumstances. These circumstances relate to Ofwat's narrow brief, which produced an active and high-functioning but ultimately dysfunctional regulator. In its regulatory brief, Ofwat was given an economics mission which involved balancing two variables: to find the lowest household bills that would generate the revenue to deliver market-acceptable investor returns. In practice, in an asset-heavy industry with a regressive charging system and little scope for cost reduction, Ofwat had a business trilemma choice. This involved choosing two out of three positive outcomes: low household bills, adequate physical investment or acceptable market returns. The current predicament of the privatised water industry and its prospects has been determined by how Ofwat did not recognise the trilemma, and in the 2000s and 2010s chose to avoid provoking customers and to keep financial investors happy, while sacrificing physical investment. The unintended consequence was that Ofwat politically protected the privatisation project for twenty years by keeping bills low and allowing investors to financially engineer higher returns. The denouement is, of course, the current industry crisis of financial and physical unsustainability.

The oversimplification of the economic regulator's choice has its origin in the 1980s, when the privatisation of state monopolies was being planned and executed, displacing government control via the boards and corporations that operated the publicly owned utility services. The result was an innovation in governance with the creation of a new kind of regulator for privatised monopolies. Ministerial powers were delegated to technocrats, including 'economic regulators'. Politicians and economists shared the naïve aim of taking the politics out of controlling privatised monopolies, though choices regarding prices, physical investment and financial returns were of course inherently political. But the new regime did secure delegation so that government ministers were not responsible for decisions on prices and other sensitive matters. Thus, Ofwat as economic regulator has operating independence to set prices for five-year periods so as to protect consumers and at the same time ensure that water companies with regional monopolies can secure an adequate return but do not make excessive profits. Since 2022, Ofwat has also been instructed to pay attention to system resilience and environmental objectives,[34] though compliance with environmental regulation is primarily the responsibility of the Environment

Failure of political and regulatory control

Agency, which does not in any way engage with industry financials.

The economists, who devised a new generic regulatory regime for water and other privatised utilities such as telecoms in the 1980s, approached their task using standard micro-economic theory of firm behaviour in uncompetitive product markets. The assumption is that, left to themselves, firms with monopoly power will raise prices and earn excess profits, without being under any market pressure to increase efficiency or service quality. Hence the objective of balancing consumer and investor interests. Regulators of newly privatised utilities added a dynamic refinement to simulate the forces of competition that are assumed to drive efficiency gains. Under the 'RPI – X' formula, firms could not increase their prices by more than general price inflation (measured by the retail price index), minus a value 'X', which was fixed by the industry regulator over a set period. If allowable price increases are set at less than the rate of inflation, this provides the incentive to the utility companies to find efficiencies. In the case of the water industry, the adjustment factor was called 'K': this could be positive or negative, depending on the expected costs of meeting environmental or other quality improvements and the view of the potential for efficiency cost savings.[35]

Operationalising this apparently simple framework has produced ever-increasing complexity. At every five-year decision point for each company, the regulators must estimate the financing cost of equity and debt capital and calculate the 'weighted average cost of capital'; consider the level of debt in the industry and calculate the 'gearing' or 'leverage'; calculate the value of the physical (tangible) assets in the 'regulated asset base'; and run a model of allowable costs that currently includes operating and capital expenditure (collectively known as 'totex'). The ability to manoeuvre around such complexities is increasingly a key competence for water companies because there has been a kind of arms race between the regulator and the companies who challenge and game the system in successive price reviews. As Dieter Helm notes, 'each price review has added new complexities and mechanisms (roughly two per price review)', increasing the scope for gaming an adversarial system.[36] In the PR19 process which set investment and prices for 2020–25, three companies thought it worthwhile to challenge their determinations and asked for a referral to the Competition and Markets Authority (CMA).[37] Following the most recent determination, PR24, six water companies – serving around half of English households – have challenged Ofwat's decision on price rises,[38] despite allowed increases in charges averaging 36% by 2030.[39]

In 2024 Water UK, the industry trade body, argued that investment was being held back by slow and inflexible regulatory processes, and criticised Ofwat for creating a 'labyrinthine framework of intense complexity'.[40] The fundamental problem is not the elaboration of the model but the inherent limits of the 1980s generic, economics-based model of regulation for privatised industries. The model did not recognise the specifics of the asset-intensity of water and its requirement for high levels of physical investment in tangible assets; nor did it recognise the extreme difficulty of finding cost reductions in asset-heavy and labour-light water operations, as described in Chapter 2.1. The generic model assumed that the regulator's task was to manage the two variable trade-offs between prices and profits created by monopoly power, so as to find the lowest customer prices which allowed adequate returns. In practice, under the existing charging system, the water regulator's task was a trilemma because Ofwat had to choose two out of three goals: low

Failure of political and regulatory control

household bills, high levels of investment in tangible assets or market-acceptable investor returns. In the 1990s the trilemma was managed by allowing bill increases; in the 2000s and 2010s it was managed by accepting completely inadequate levels of investment.

Ofwat has not recognised the intractability of the trilemma. Balance was the theme in Ofwat's written evidence to a 2018 parliamentary committee, when it explained: 'we regulate to ensure ... resilient, reliable, and high-quality services to customers in an efficient way balancing the need for continued investment with bills which customers can afford and are willing to pay'.[41] In the same submission, Ofwat boasted that average bills were in real terms much the same as twenty years previously, and that PR14 had delivered real bill reductions over the period 2015–20; all without pausing to consider whether this was because physical investment levels had been grossly inadequate. As for mainstream economists, after more than thirty years some have belatedly recognised that the idea of taking the politics out of regulation was flawed and is unworkable. In oral evidence to the Lords Select Committee in 2022, Catherine Waddams (an academic and previous member of Ofwat's board) acknowledged the need for decisions on 'trade-offs ... between different objectives', and went on to say, 'I would argue that should sit with the Government or Parliament, not with the regulator'.[42] Political instruction, of course, would not remove the need for choice and make the trilemma any easier to manage, though it would make the choices more visible.

If Ofwat resolved the trilemma for twenty years by privileging low bills, this reflected a very narrow definition of the public interest in water and sewerage. Ofwat's repeated use of the term 'customer' is itself significant here. This reflects the fact that Ofwat is a very small agency, currently employing no more than about 250 people,[43] which continues to draw overwhelmingly on economics-based expertise. The ultimate beneficiary of regulation is not constructed as a citizen or even a consumer who may also have environmental interests which extend to outcomes for future generations. The beneficiary is constructed in a narrow economics framework as a customer who pays for a specific good or service and is concerned with quality and price here and now. As late as 2018, in written evidence to the House of Commons Environment, Food and Rural Affairs Committee, Ofwat insisted that it wanted 'water companies to put customer interests at the heart of their business'.[44] Ofwat had no broader definition of *citizen* interests until it was politically instructed by the government in 2022 that its strategic priorities should be to 'protect and enhance the environment' and 'deliver a resilient water system', as well as to 'serve and protect customers' and 'use markets to deliver for customers'.[45]

In the first decade after privatisation this narrow definition of customer interests was reinforced at Ofwat by the sidelining of environmental issues. The dominant figure in the regulation of water in the 1990s was Ian Byatt, an ex-Treasury civil servant who was director general of Ofwat from 1989 to 2000. In a 2012 memoir, Byatt recalled how he had fought off Lord Crickhowell at the National Rivers Authority (an earlier regulator that was subsumed into the Environment Agency) who wanted 'quality enhancement'. Byatt wanted economic cost–benefit analysis applied to ensure affordable bills that would 'increase by no more than inflation, or the growth of household income'[46] because 'customers need protection not only from monopoly water companies seeking high returns but also from environmental

Failure of political and regulatory control

groups including government departments pushing for water quality and environmental enhancement'.[47]

By the 2020s this kind of reactionary view was no longer acceptable (at least in public). On its website, Ofwat adopts the language of long-term social responsibility. The regulator's strategy now includes driving water companies 'to meet long term challenges', while its vision includes 'long term stewardship of the environment', with duties encompassing securing 'the long term resilience of water companies'. The problem is that Ofwat might talk about the long term, but it regulates companies within a short-term framework set by the five-year price review process. Acting long term would require developed national and catchment plans for large-scale investment over twenty years and a pipeline of projects including nature-based projects designed to deal with the challenges of the next fifty years. What we had in 2024 was Ofwat and the water companies negotiating around a doubling of grossly inadequate investment for the years 2025 to 2030. At the same time, as we detail in Chapter 6.2, Ofwat was on the side approving ad hoc company proposals for steel and concrete PFI projects such as reservoirs and sewage treatment works, which would not be on water company balance sheets. This should raise the question, if major projects are now dependent on PFI, what was the point of the regional water companies?

While economists were worrying about the problem of monopoly power in product markets, they did not notice the conjunctural development of financialisation which brought new kinds of investors demanding higher returns for a few equity holders, even when this undermined the sustainability of an operating business in the future. Financialised management had gone mainstream in the UK in the 1980s with Hanson PLC,

Murky water

an acquisition-driven conglomerate. In the 1990s financialisation was institutionalised by private equity investors using debt finance and leverage to boost equity returns from operating companies.[48] By the end of the decade, the shareholder value movement had enforced financial performance targets for all US and UK public companies, even in asset-intensive utilities such as water.[49] The 2000s saw the emergence of behemoths such as Blackstone and Macquarie, which were both investment managers and principal investors in asset classes such as infrastructure, which also attracted fund investors such as the Middle East and Far Eastern sovereign wealth funds.

This meant that, while Ofwat was doing its narrow regulatory job of attempting to curb the power of monopoly suppliers to ramp up profit margins, it did nothing to stop financialised investors who were extracting cash and wrecking balance sheets by loading them with debt. For financialised investors, the challenge is how to extract above-average returns on investment from mundane businesses, where profit rates are limited by various combinations of competition, capital-intensity or regulation. Put simply, the objective of financial engineering is to extract cash for shareholding owners from an operating business, typically over a period of around five to seven years. If there is little scope for operating cost reduction (as in the case of water), cash extraction can be increased by one or more of three classic devices: restrict capital investment, sell assets or load the operating business with debt. As regulated water assets cannot easily be sold (unlike retail or warehouse property), cash extraction in water has to operate by some combination of restricting physical investment and taking out debt. Increasing borrowing while shifting the funding mix from equity towards more fixed interest debt created the possibility of higher equity

Failure of political and regulatory control

returns. This was the case with Thames Water where, as we saw in Chapter 2.3, the company went from being around 15% debt-financed in 1990 to 90% by 2013. These financial extraction techniques may weaken the long-term sustainability of the business but, as financialised investors are generally not long-term owners, the necessary restructuring can be dealt with by the next owner.

Financialisation is not only about processes of cash extraction but about corporate structures which are tax-efficient and opaque, so they are more difficult for outsiders to understand and control. As Ofwat was regulating the operating company (which includes the regulated assets), there were opportunities for financial engineers to find advantage and obscure what was going on by creating complicated multi-level corporate structures. To illustrate this, Exhibit 5.1 summarises the ownership structure of Thames Water, where in 2023 there were six layers of companies, from the holding company, Kemble Water, to the operating company, Thames Water Utilities. This structure creates all kinds of possibilities for cash transfer via interest and dividend payments in different directions through multiple levels. For example, in late 2023 Ofwat queried breach of licence conditions when Thames Water paid a £37.5 million dividend from its regulated operating company to 'service external debt obligations' of its parent holding company and one of its subsidiaries.[50] Thames Water replied that it was licence-compliant because strictly speaking this was not an 'external dividend'.[51] This illustrates how complex corporate structures, which are entirely unnecessary for the provision of water and sewerage services, make it very difficult to 'follow the money'.

The financial engineering does not stop at complicated ownership structures, because many water companies make

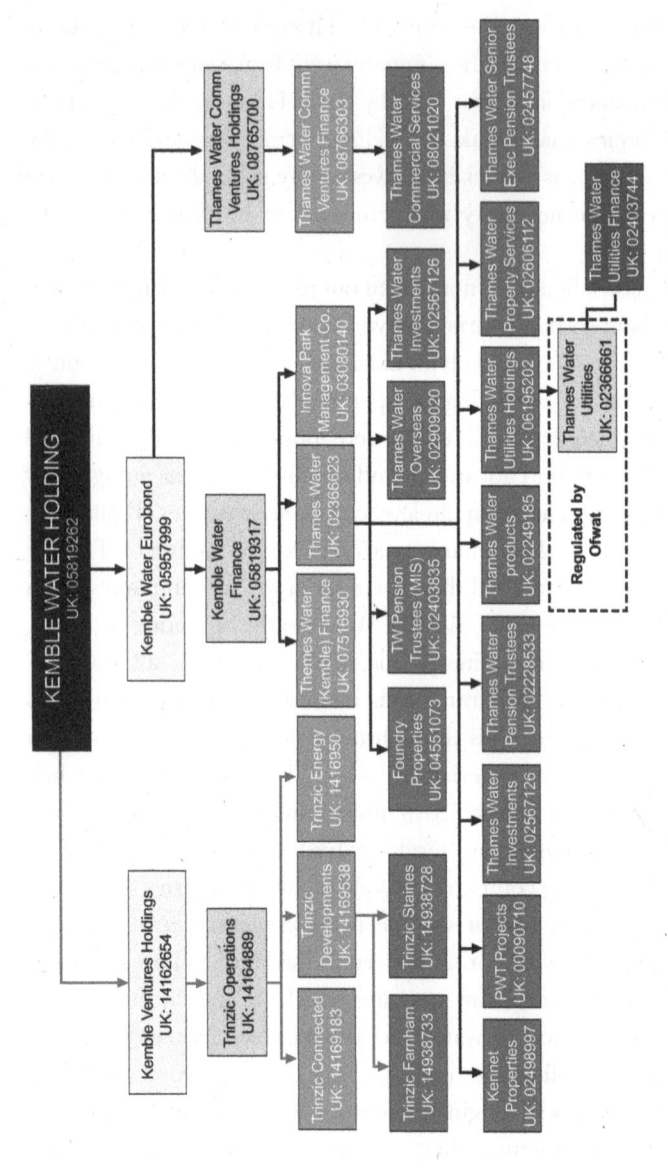

Exhibit 5.1 Corporate structure of Kemble Water Holdings, owner of Thames Water, 2022/23[52]

Failure of political and regulatory control

extensive use of complex financial products that can be a source of trading gains or losses for the operating business. These financial products are typically used to hedge against currency risk or changes in commodity prices. In the sheltered activity of water, derivatives are being widely used to hedge against fluctuations in interest rates, which are of course important in debt-financed companies. In 2021 Yorkshire Water had a total index-linked swap portfolio of £1.29 billion, with mandatory breaks between 5 and 17 years to limit concentration risks.[53] As with any hedge, derivatives can increase or decrease the cost of servicing Yorkshire Water's debt, with different outcomes in various years. The water companies claim to be behaving prudently here, but it is very difficult to understand what they are doing from their financial accounts, or to see in any one year whether their debt-service costs include gains and losses on financial trading.[54] It is also not clear that Ofwat understands the risk profile of derivatives portfolios.

Ofwat did not register financial engineering and the growing debt mountain before their existence was widely advertised by Kate Bayliss and David Hall, as discussed in Chapter 2.1.[55] By 2018 everybody, including the Secretary of State Michael Gove, understood what was going on, but Ofwat still failed to control financial engineering in its PR19 price determinations. The problem by then was that Ofwat had updated its economics and was engaging with financialisation through a naïve understanding of corporate finance theory. In the Modigliani–Miller theorem, the mix of equity and debt finance has no implications for the valuation of a company, in a perfect market without taxes or financing costs. As debt is cheaper and tax advantageous – because corporation tax is levied after interest payments while dividends are paid from post-tax profit – the expectation is that gearing up with a mix of more debt and less equity is an efficient way of financing a company. In the 2010s interest rates were low and Ofwat was overly relaxed about the growth of debt and the increased ratio of debt to equity in water companies. In its key position paper before PR19, Ofwat did not try to cap debt and gearing ratios but bizarrely complained that all the benefits of cheap debt financing were being captured by investors, not lowering bills.[56] Ofwat proposed that in highly geared companies these benefits should be shared with customers through a 'benefit sharing mechanism to better align the incentives of companies and their investors around choice of gearing levels with customer interests'.[57]

As for dividends to shareholders, Ofwat maintained a policy of non-intervention even after the problem of excessive distribution of profits had been uncovered by Bayliss and Hall. In its key position paper for PR19, Ofwat explained that it would ask water companies to explain their dividend distribution decisions, but it would not seek to cap payments.

> Our approach is to regulate prices charged to customers and to set allowed returns for investors, as part of setting price controls. We do not regulate the level of dividends. We recognise that decisions as to the declaration and payment of dividends are best determined by companies and their boards, within the wider framework of price controls, licence obligations and company law.[58]

This hands-off approach consolidated Ofwat's failure in PR19 to act on public knowledge about financial engineering and restrict cash extraction in all its forms. The resulting burden of debt on wrecked balance sheets stored up sustainability problems and the problems of high debt have continued to worsen. But Ofwat has remained remarkably complacent about its ability

Failure of political and regulatory control

to provide effective oversight of the sector. In evidence to the House of Lords Industry and Regulators Committee inquiry into Ofwat in 2022, the CEO of Ofwat was asked why Macquarie was allowed to buy Southern Water after having been 'a highly unsatisfactory owner of Thames Water'. He replied:

> The reason why we were not uncomfortable with the proposal was that we have changed the regime since Macquarie owned Thames. We think we have built additional protections into the regime to stop the kind of behaviour that we saw at Thames and the poor performance there.[59]

Only two years later Southern Water posted its largest ever loss, had its credit rating downgraded, putting it at risk of default due to bond covenants, and was on Ofwat's 'financially at-risk list'.[60] This debacle strongly suggests that Ofwat's regulatory 'regime' remains inadequate to the accounting task of regulating financially engineered businesses. In Chapter 6.2 we see that Ofwat is now effectively licensing the large-scale use of PFI-type schemes to bring in new finance for major projects. Given Ofwat's long record of failure to understand or control financial engineering, this is an alarming development. Regulatory reorganisation and the scrapping of Ofwat's 1980s economics balancing brief is long overdue, though this will only produce benefits if it broadens Ofwat's narrow expertise base and reinvents the economics regulator as a financialised business regulator.

5.3 The Environment Agency: sponsored regulation

The problems with economic regulation arise from both the way that Ofwat's object was very narrowly defined and from its large measure of operating independence. All this is turned upside down in environmental regulation. The Environment Agency is a multi-agency, created by a 1995 Act which merged many smaller agencies (including the National Rivers Authority, established at the time of privatisation) to create one organisation in England with multifarious responsibilities 'to protect and improve the environment' in water, land and air, with its counterpart, Natural Resources Wales.[61] Thus, apart from its regulatory role in the water industry, the Environment Agency is responsible for many other things including flood risk management and the regulation of industrial pollution and business waste. The Environment Agency is by UK standards a large agency, which currently employs around 12,500 people, but is not adequately resourced to fulfil its multifarious duties.[62] Equally the Agency is not empowered to operate independently of political control because it is formally 'sponsored' by Defra, which is responsible for the environmental policy framework within which the Agency must work. The Environment Agency is, then, an ineffectual, low-functioning regulator which is politically useful because it can provide cover for ministers facing criticism on environmental issues.

In the water and sewerage industry, the Environment Agency has a high-profile, public-facing role. It is responsible for maintaining and improving the quality of rivers, lakes and coastal waters through the inspection of treatment facilities, monitoring of sewage discharges and enforcement response to pollution incidents. But the Agency also has many other important operating responsibilities in the water industry, which include the licensing of water abstraction for public water supply or commercial use in industry and agriculture. With the proliferation of documents such as Defra's 2023 *Plan for Water*, in recent years

Failure of political and regulatory control

the Environment Agency has been given long-term strategic responsibility to secure the proper, efficient use of water resources in England. Thus, the Agency now leads on the development of the Water Industry National Environment Programme and is responsible for assessing the water resource plans of each water company.[63]

The field of environmental regulation has recently been complicated by the way in which the economic regulator Ofwat has also gained some environmental responsibilities in an attempt at strategic thinking and joined-up government. As we have already noted, Defra's most recent Strategic Policy Statement in 2022 laid down supplementary guidance for Ofwat. In a way that 'complements Ofwat's existing duties', the regulator should now 'prioritise appropriate action to enhance water quality and deliver a resilient and sustainable water supply'. Specifically, Defra directed Ofwat towards 'progressive reductions' in storm overflow discharges. It has also instructed Ofwat to weight long-term interests by 'recognising that a system that works in the enduring interests of consumers does not simply mean lower prices in the short-term at the expense of future generations',[64] though it is unclear how this should sit alongside its other responsibilities, nor how Ofwat will work effectively with the Environment Agency.

For much of the 2010s the Environment Agency was effectively not an independent regulatory agency but a way of providing political cover for Defra and successive ministers when inquiries were being made, explanations had to be given, and criticism had to be fielded. This much is clear from the location of the Environment Agency press office in a Defra group media office.[65] Agency executives can be called in front of parliamentary select committees or front up for the media whenever it is necessary to explain operations, admit inadequacies and promise to do better. When new horrors are uncovered, government ministers can be suitably shocked or puzzled before going on to say that it is the Environment Agency that is the regulator. Here, for example, in 2022 is the then Defra Secretary of State Therese Coffey in 'nothing to do with me' mode. She was responding to a BBC investigation which had uncovered dry spills from storm overflows without the excuse of rainfall. 'It does seem extraordinary on the hottest day of the year that there may be releases. The Environment Agency is the regulator; they are the people who do the detailed investigation of why that has happened.'[66]

For much of the period since the early 2010s the under-resourced and ineffectual Environment Agency was effectively operating on the principle of *see no evil, hear no evil and speak no evil* about a seriously delinquent water industry. This performance is epitomised by the charade of the Agency's Environmental Performance Assessment.[67] Year-by-year since 2011, this has graded water companies according to various key performance indicators, complete with a star rating system and league table that serves to sustain the illusion of a responsible, functioning industry seeking improvement. At the bottom end, the table identifies four laggard companies (Thames, Southern, Anglian and Yorkshire) as being responsible for more than 90% of serious pollution incidents in 2023.[68] But in that year United Utilities regained its four-star status which represents 'industry leading performance', even though in the same year it had been pilloried for irresponsibility in a BBC *Panorama* documentary on pollution in Windermere.[69]

The backstory is that the Environment Agency has acquiesced in its own ineffectuality. There were step change reductions in compliance activities in 2009 and 2015 after the introduction

Failure of political and regulatory control

of the operator self-monitoring (OSM) system, which since 2010 has allowed water companies to self-report pollution incidents. The entirely predictable result is under-reporting by companies. In 2022, for example, only 48% of serious pollution incidents (category 1 and 2) were self-reported by the companies.[70] The 2023 *Panorama* report on United Utilities documented how the Environment Agency attends few pollution incidents and allows self-reporting whereby category 2 and 3 incidents are downgraded to less serious category 4, 'with no environmental impact'.[71] The OSM system has been completely discredited as a result of civil society campaigning and media investigations, and even the trade body Water UK has asked that it be replaced.[72] In 2021 the then water minister announced that the OSM system of incident reporting would be replaced by independent inspection, and this remains government policy. However, the system of self-monitoring limps on, with no firm deadline for its complete scrapping. In January 2024 a Conservative minister grandly announced that 'the era of self-monitoring is over';[73] but in November 2024 a Labour minister explained that this meant that the Environment Agency was in process of recruiting more inspectors to carry out independent audits, 'reducing the reliance on operator self-monitoring'.[74]

The monitoring of water quality has never been adequate, and it was further undermined at the end of the 2010s. In 2023 the Environment Agency admitted that the monitoring programme 'has been cut back considerably in recent years', and the Agency was then doing 'about a third of the Water Framework Directive monitoring checks that we were doing four years ago'.[75] The Agency was also not monitoring a series of other important issues such as water abstraction. Since 2017 there have been plans for the reform of abstraction, which is taking place under long licences without volume limits,[76] and in 2022 charges were raised for those (such as water companies) who were abstracting large quantities.[77] However, there is no clear monitoring and inspection regime, even though in 2018 the Environment Agency admitted that 'current levels of abstraction are unsustainable in more than a quarter of groundwater bodies and up to one-fifth of surface waters, reducing water levels and damaging wildlife'.[78] We explore this further in the appendix to this chapter.

Ineffectuality is also manifest in the way that Environment Agency priorities are not being set proactively and internally, as they would be in a technocratic set-up. Priorities appear to be set in response to external events, ministerial priorities and public concerns, as they are in politics. The Environment Agency is formally organised into different divisions, but from oral evidence to parliamentary committees, senior management appears not have the bandwidth to actively deal with multiple, unrelated, complex problems.[79] For example, in the mid-2010s, after the exceptional rainfall and winter floods of 2015–16, the concern was with the need for a 'complete rethink' of flood defences and flood warnings.[80] By the early 2020s public indignation about storm overflows could not be avoided, and the Environment Agency and Ofwat in response mode had to be seen to be doing something. This brought about two major changes: first, environmentally underperforming water companies could now expect punishment in the form of large fines; second, after all storm overflows had been fitted with monitors, the public would have access to hard data on discharges, published for the first time in 2025.

Ofwat has discovered performativity and has come out swinging with fines for environmental offences. The Environment

Failure of political and regulatory control

Agency had routinely fined companies for offences, but the amounts levied were usually token (in relation to company turnover), with fines typically around £1 million per offence. Between 2015 and 2023 the Environment Agency secured 63 prosecutions against water companies which altogether raised not much more than £151 million,[81] with £90 million of that total accounted for by one large fine against Southern Water for gross offences compounded by deception.[82] In 2024 Ofwat fined three companies (Thames, Yorkshire and Northumberland) £168 million for excessive spills from storm overflows caused by poor operation and maintenance of treatment facilities and by failure to upgrade capacity.[83] At the same time, Ofwat announced that eight further companies were under investigation and, by implication, that most companies could expect hefty fines.

There can be no doubt about Ofwat's willingness to levy substantial fines which formally cannot be recovered from customers. Ofwat can fine companies up to 10% of sales revenues and has already fined Thames and Yorkshire at rates of 9% and 7%, though in 2025 Thames was reportedly requesting that current and future fines be waived to help it find a new buyer.[84] The Environment Agency has separately gained new powers to issue unlimited financial penalties without criminal prosecution.[85] However, exemplary financial punishment does not prevent environmental harm, and the issue is then who funds and pays for remediation. For example, sewage spills are caused by under-capacity and patching repairs so that companies will have to invest substantially in upgrading and replacement of equipment in dilapidated treatment works, and in many cases find investment funds for extra capacity. Debt-burdened firms with wrecked balance sheets such as Thames have limited capacity to finance enhancement investment by borrowing more, hence the importance of what we describe in Chapter 6.2 as projectification. Major projects will now be funded outside of water company balance sheets by financial consortia under PFI-type deals whose cost will be added in surcharges to customer bills. If levying fines to encourage physical investment is the easy bit, controlling projectification is far more difficult.

To address the high-profile problem of untreated sewage discharges, the fitting of storm overflow monitors is obvious and long overdue, but, like punitive fines, it promises more than it delivers. For twenty-five years after privatisation, 95% of storm overflows were not monitored. In 2013 the Environment Agency gave the water companies a leisurely seven years to fit event duration monitors (EDMs), which measure when and for how long a storm overflow is in operation, to most of their storm overflows. And, on National Audit Office calculations, as late as 2016 only 6% of storm overflows were monitored and reported to the Environment Agency.[86] At the end of 2023, however, Defra triumphantly announced that all 15,000 storm overflows had been fitted with EDMs.[87] In spring 2024 the Environment Agency reported that the average number of spills per overflow was 33 and in total there had been 3.6 million hours of spills in 2023.[88]

The issue here is that reporting of spill events encourages 'shock, horror' reactions without establishing the context in ways that allow focused remedial action. Time series analysis of trends and an understanding of year-on-year variation according to rainfall is impossible because most overflows were not monitored until very recently. The absence of time series evidence also makes it difficult to establish a justifiable benchmark against which we could set and assess industry targets for spill reduction. The other issue is the lack of monitoring data that relates spills

Failure of political and regulatory control

at a specific site to local rainfall. The Environment Agency is working on mapping discharges against rainfall, but in 2024 comprehensive data was described as an 'aspiration'.[89] It is therefore impossible to decide which spills are reasonable – because they are excused by heavy rain[90] – and to identify what percentage of spills for each company are completely irresponsible 'dry spills'.

The absence of system-wide evidence on dry spills is alarming. Dry spills are unlawful for good environmental reasons because they tip undiluted sewage into waterways. Repeated dry spills also indicate egregious corporate irresponsibility because they reflect some combination of lack of hydraulic capacity to treat routine flows, inadequate plant maintenance, or even an operating decision to avoid the expense of sewage treatment. The BBC, through a Freedom of Information request, found out that in 2022 three companies (Thames, Southern and Wessex) illegally released sewage 388 times for a total of 3,572 hours on dry days.[91] BBC reporters tried to cross-reference their data with Environment Agency records on illegal dry spills and found that the Agency's records covered only a third of the cases uncovered in their investigation.

More broadly, facilities inspection, water quality monitoring and pollution incident response by the Environment Agency are only in the early stages of recovery from complete collapse. As the Agency admitted to the House of Lords Industry and Regulators Committee in 2023, 'it has not had the necessary funding or data access to take sufficient enforcement action against water companies'.[92] Effective inspection requires inspectors, evidence and enforcement powers, and it is clear that the Environment Agency has not had the resources for regular and unannounced inspections of every sewerage works. Through the 2010s the

Agency had the target of visiting every sewerage works at least once every eight years, but it was actually visiting no more than 6% of sites each year, which corresponds to once every seventeen years, and after 2019 it had no target for regular inspection of every works.[93] In 2023 the Agency admitted that it had only 91 qualified inspectors who in one year had carried out 1,420 inspections (of 16,000 regulated installations in water and sewerage). Belatedly, the Environment Agency is promising to do better and appoint 500 qualified inspectors to carry out 10,000 inspections in 2025,[94] and funding has been given to carry out more inspections.[95]

The commitment to improve the inspection and enforcement of sewage treatment is a welcome development. However, the Environment Agency is moving slowly on too narrow a front to address water quality issues caused by agricultural and road runoffs as well as sewage. The issue here is not simply one of under-resourcing but of ministerial sponsorship and the lack of regulatory independence which would allow the Agency to be politically awkward. All this is compounded by confusions in the Agency's sponsoring ministry. As we have noted, Defra gets diverted into issue management. The National Audit Office has also reported that the department lacks a clear understanding of how it can meet the government's environmental goals[96] and notes that it has not worked effectively with its regulators.[97] Against a background of regulatory underperformance, the managerialist solution would be to create one unified water regulation agency by merging Ofwat and the relevant parts of the Environment Agency with the Drinking Water Inspectorate, the third regulator of the water industry. There is a case for merging business and environmental regulation of water, but this would only deliver benefits if the restructuring combined

Failure of political and regulatory control

new expertise for business regulators; independence for adequately resourced environmental regulators; and strong political support from Defra, rather than issue management. The rearrangement of regulation in itself will not solve the problems of political direction which go well beyond issue management, as we analyse in the next chapter.

Appendix: avoiding abstraction

The three stories that opened this book frame public knowledge of water and sewerage, and they also limit the field of the visible and set priorities for action. This is particularly so in the case of the two affective stories – financial extraction and failure of service – because 'fat cats' and sewage in rivers feed public indignation. This distracts from the equally serious issue of how the water supply in the south and east of England depends on environmentally irresponsible abstraction of surface water from rivers, and of groundwater, especially from chalk aquifers. Both reduce river flow, with adverse consequences for biodiversity, including declines in trout and grayling stocks and threats to wading birds that rely on intertidal habitats. But ecology is often crowded out in media stories or sidelined in media coverage. For example, a 2025 BBC documentary on Thames Water featured graphic coverage of sewage spills from decrepit works with inadequate capacity such as Mogden. At no point did this documentary mention that the London and Thames Valley drinking water supply comes from abstraction: London's supply is 80% from surface water and 20% from groundwater, while in the Thames Valley as a whole supply is 30% surface water and 70% groundwater.[98]

Murky water

The National Audit Office explains that abstraction through inlets or boreholes is 'the cheapest way for water companies to source water' because, as in London, it requires only small storage reservoirs to buffer supply.[99] Hence, half of English non-tidal surface and groundwater abstraction is for public water supply, and abstraction dominates in the south and east of England where topography and geology make it cheap and easy. Topography favours surface water abstraction. The rivers Thames and Lea (which joins the Thames at Bow Creek) have historically been the source of London's water, and currently the Thames provides two-thirds of London's drinking water. More broadly, geology is favourable across most of the south and east of England, which has around 200 chalk streams where the water-bearing aquifer can be tapped. According to the British Geological Survey, groundwater accounts for 75% of water supply in south-east England,[100] so that, for example, Cambridge's water supply comes from chalk aquifer tapped by 24 bore holes.[101]

Much of this abstraction is environmentally irresponsible. Water companies, government and regulators have for twenty years known about the problem of over-abstraction which exceeds aquifer recharge rates and reduces river flow, thus compounding the water quality problems created by agricultural runoffs and sewage discharges. On Defra's calculation in 2023, 15% of English rivers and 27% of groundwaters were 'over-abstracted'.[102] Pressure on low flows is regionally concentrated in the dry south-east of England where the population is increasing. The South East Rivers Trust cites World Resources Institute research which suggests that a 'warning threshold' is reached when about 20% of freshwater resources are abstracted, and an area is classified as 'severely water stressed' at the 40% level. This analysis found that 52% of available resources are being abstracted in

Failure of political and regulatory control

the Thames River basin, and 35% across the whole of the south-east.[103] The water companies are abstracting every day of the year at the expense of recharge and flow rates. Environment Agency abstraction data shows that in most of the east and south-east of England water can only be 'taken sustainably from the environment' on less than 30% of the days in the year, and in much of the rest on less than 50% of the days.[104] Limiting abstraction to 10% of annual average aquifer recharge to improve chalk stream health would require immediate abstraction reductions of 610 million litres per day.[105]

All this is now an increasingly urgent problem which requires immediate reductions in abstraction volumes, because England and Wales are well short of meeting targets on water quality, and climate change is aggravating the consequences of over-abstraction. After Brexit, Defra and the Environment Agency carried over the EU's Water Framework Directive requirement that all surface water bodies, including rivers, should achieve 'good ecological status' by 2027 (or good ecological potential in the case of 'heavily modified' water courses). The OEP in 2024 found that this target was going to be missed by a large margin and, in its worst-case assessment, only 21% of surface water bodies would meet the standard.[106] The consequences of over-abstraction become increasingly dire with climate change, which brings increased temperatures and modified rainfall patterns. The Environment Agency is projecting an average drop in river flows of 15% by 2050 due to climate change,[107] which means that abstraction needs to be reduced significantly to support improved river quality and aquifer recharge.

Policymakers have responded in a very ambiguous way. They have set responsible, ambitious targets for abstraction reduction by 2.8 billion litres a day.[108] But that target is embedded in a broader strategy for meeting a 5 billion litres a day projected shortfall in water supply by 2050. This strategy depends on implausible assumptions about lower per-person consumption and fewer mains leaks which reduce demand for water, as explored in Chapter 3.3. If water usage is not reduced – and given that new build of reservoirs and water transfer is expensive and slow – the pressures for continued high levels of abstraction will be irresistible.

- The 2024 summary of England's water resource management plans project 'abstraction licence reductions' of 2.8 billion litres per day by 2050, 'largely to address existing unsustainable abstraction practices'.[109] This translates into abstraction reductions of more than 10%, required not only in the south-east but also in the west midlands and the north-west. At least 30% reductions are required across large areas of the groundwater-dependent south-east and the surface water-dependent west midlands.[110] As the chief executive of the Environment Agency told the Environmental Audit Committee in 2024, this nearly 3 billion litres per day reduction in levels of abstraction is the largest single contributor to an estimated 5 billion litre shortfall in water supply by 2050.[111]
- In turn, that shortfall estimate is dependent on assumptions about demand reduction by 2050, specifically a 50% reduction in mains leakage and a 25% reduction in personal consumption.[112] As Chapter 3.3 argued, these assumptions are implausible because they involve a break with past trends, which is empirically unjustified but politically convenient because it reduces the requirement for highly expensive new construction. If and when these demand reductions

Failure of political and regulatory control

do not materialise, with population increase and business development continuing, there will be irresistible pressure to reduce the supply shortfall by dropping the responsible abstraction reduction targets and continuing with irresponsibly high levels of abstraction.

We can have little confidence in the ability of Defra and the regulators to manage these incoherences, given that they have a twenty-year record of worthy ambition not matched by delivery. Indeed, up to about 2018, the record on abstraction is of inaction, as reflected in a culpable failure to tighten conditions around abstraction licences so as to link licences with water availability. Licences are required formally for abstracting over 20 cubic metres, and the water companies together hold 1,400 licences, often allowing extraction up to a fixed volume per day or rolling period. In 2017 the Environment Agency chief executive told a House of Commons committee that 'many of the abstraction licences have no limits on the amount of water that you can take out of the ground, and many of those abstraction licences are very long-term'.[113] Ofwat in a 2015 report had also complained about licences not being linked to water availability,[114] with abstractors required to annually self-report the quantity abstracted, allowing for no real-time monitoring or adjustment.[115] The fining of United Utilities in 2023 for exceeding a rolling limit on groundwater abstraction was newsworthy because prosecutions for breach of licence are highly unusual.[116] Since 2003 new licences have been subject to renewal, although in 2018, 70% of total licences (accounting for 50% of all water abstracted) were still permanent.[117]

The overwhelming impression in the 2000s and 2010s is of dilatory and negligent regulation and governance, despite the

2027 water quality targets. This impression is reinforced if we remember that public water supply accounted for 51% of non-tidal surface and groundwater abstractions in 2018, meaning that around half of abstraction was by other users.[118] Defra and the Environment Agency know almost nothing about industrial and agricultural abstraction and, as late as 2020, the NAO was complaining that industrial and agricultural uses were not included in water company resource forecasts.[119] It must still be doubtful whether the current official water demand forecasts include informed estimates of the increased agricultural demand for irrigation with climate change.

There has in the 2020s been a flurry of regulatory reform. This has involved new powers, as in the Environment Act 2021 which allows the Secretary of State to vary or revoke permanent abstraction licences without compensation where there is a risk of 'serious damage' to the environment.[120] From 2028 this can also be done if 1) it is necessary having regard to a relevant environmental objective; or 2) to otherwise protect the water environment from damage.[121] This extension is potentially valuable but it requires sufficient regulatory resource to monitor abstractors and to take the actions required to vary or revoke licences, and to enforce these changes. A Defra report in 2019 on abstraction reform made all the right noises about catchment areas and reported that four 'catchment groups' were discussing innovative solutions.[122] But it is not clear what difference pilot activity makes to outcomes when Defra has a track record of over-claiming. For example, a 2019 Defra blog claimed that 'by reviewing licences and reducing the amount of water people can take we have returned 16 billion litres of water back to chalk aquifers and streams since 2008, and removed the risk of another 14.9 billion litres being taken'.[123] These volumes are

Failure of political and regulatory control

impressive only because they are annualised, and nearly half the total saving (14.9 billion litres) is notional because it comes from revoking unused abstraction licences.

What is increasingly clear is that abstraction reform is achieving little because Defra and the Environment Agency are not working back from the objective of 'good ecological outcomes' to a specific programme of measures for each water body which address excessive abstraction, sewage spills and agricultural runoff, which, in different combinations, degrade water flow and quality. This is the only sensible way of tackling interrelated problems. Indeed, Defra and the Environment Agency are prepared to fight against this approach all the way to the supreme court, as they did in the Costa Beck case brought by Fish Legal and Pickering Fishery Association. This case was not about abstraction because the Costa Beck pollution was caused by sewage and fish farming, but it did establish a guiding principle which is relevant to all sources of degraded water quality. Defra and Environment Agency lawyers unsuccessfully argued that the high-level and generic Humber River Basin Management Plan was an adequate guide to action. Fish Legal won a precedent-setting judgement that environmental objectives could only be met by a programme of measures which set out specifics for each water body to reach good ecological status.[124]

By the measure of outcomes, we have made little progress in reducing damaging abstraction in a decade. In 2025 we could sum up the position much as the World Wildlife Foundation did a decade earlier in its 2015 report on chalk streams:

> Significantly, we're well aware of the problems and we know what needs to be done to reduce the huge pressures on our most vulnerable rivers and streams. The trouble is the improvements to date have been too little, too niche, too slow. What we urgently

need now is an effective, determined effort to push forward the essential changes and significantly upscale the solutions already being trialled in chalk streams across the country.[125]

What we have learned is that the official political apparatus of regulation and governance will deliver very little effective effort for essential changes and that outcomes depend on civil society mobilisation. This is a central theme of our final chapter, and it is relevant to a water course near you when, as we shall argue, private finance projects are allowing companies to propose investment projects which solve their problems but are not clearly in the public interest. Thus, Thames Water has the Teddington Direct River Abstraction drought-resilience scheme, 'to abstract up to 75 million litres of water each day during periods of prolonged dry weather'[126] from the Thames above Teddington weir, and then to maintain river flow by replacing the water with treated sewage carried by a 4 km tunnel from Mogden sewerage works.[127]

Chapter 6
Towards foundational water management

Introduction

In the introduction to this book, we outlined the ideal of foundational water management to deliver sustainability and liveability. Over the following five chapters we provided an analysis of how and why we have been moving, not towards this ideal, but in the wrong direction. By following three key threads – how money moves, how narratives focus and distract, and how power works – we have seen how the issues regarding what has gone wrong (and what is to be done) are complex. Having complicated the problem, in this final chapter we return to foundational water management and set out how a social movement could reorient water management towards sustainability, liveability and democracy.

In *Section 6.1* we start from what water and sewerage should provide so that citizens have the freedom to live lives they have reason to value, including living in environmental security with equitably funded clean water and wastewater treatment. Securing water management that meets foundational needs for liveability and sustainability is anything but straightforward. Hence, the need to begin by identifying a set of enabling technical reforms that could change the trajectory of the existing system and modify its tendencies. The first two of these are reform of water charging, with public ownership as a backstop against the irresponsibility of financialised owners. More ambitiously, the second two are catchment and national planning of the water and sewerage system. Planning and coordination at different scales, from management of an individual water course to inter-regional transfers of water, requires learning and experimentation, which is why foundational water management needs to be democratic and open to a diversity of expertise.

Section 6.2 highlights an immediate barrier to realising the foundational vision. The analysis here follows on from Chapter 2, which showed how the business model of the water companies leads them to under-investment and ever more debt along a trajectory towards physical and financial unsustainability. By 2025 Thames was already a zombie company, running out of cash and looking for a new equity investor and debt write-down. The broader industry-wide consequence of unsustainability is that none of the companies can raise new debt to fund major projects such as reservoirs or water treatment works. And the outcome is what we call projectification, where companies propose major projects, with each one to be financed, built and operated by one-off private finance

Towards foundational water management

vehicles. This undermines accountability and water system planning, as well as creating toll points where customers will pay through bill surcharges over anywhere from 25 to 125 years.

Zombie companies and murky projectification are indicators of how things have gone badly wrong, and the task is to change the trajectory of the industry so that it moves closer to delivering sustainability and liveability. *Section 6.3* shifts from the technical to the political and identifies the obstacle of a power configuration whereby elite groups of politicians, technocrats and corporate executives operate under the tutelage of finance in a quasi-oligarchic way. When elites agree that the water industry must be investable, the industry operates in the interests of the finance sector. Little gets built for thirty years, up to the point in the mid-2020s where projectification opens new opportunities for finance profits. When this is part of a broad and durable power configuration, citizens cannot realise more foundational water management by voting in a new government. With justified public indignation about storm overflows we have citizen campaigns for cleaner water. But these are not in themselves focused and powerful enough to block current developments such as projectification and redirect the system towards the long-term public interest.

If the obstacle to progress is a power configuration of financial, technocratic and political elites, the issue is how to disrupt this power configuration, and this is the focus of *Section 6.4*. There is a base to build on when civil society organisations have effectively highlighted pollution through

> building capabilities in citizen science as much as through campaigning. However, there is an important distinction between civil movements that largely work within existing frameworks and social movements that set out to change them. Across Europe, social movements have halted or reversed the privatisation of water and experimented with democratising water company management. In the UK there are early, encouraging signs of a developing water movement that recognises some of the broader challenges set out in this book, and could form the nucleus of a social movement that challenges power through active citizenship. The outcome depends on education and organisation.

6.1 Foundational water management and the right moves

Our starting point in this chapter is the question of what we want from our water and sewerage infrastructure. Before elaborating policy or other changes aimed at renewing the management of water, we need first to explore how to think about the necessarily long-term public interest in water. This section, therefore, explains the objectives of foundational water management as an ideal, before considering the actions or technical changes in policy and practice which are necessary if we are to move towards this vision. The aim is to establish a new kind of water management in England and Wales that is fit for purpose in an age of climate change and faces up to the multiple dysfunctionalities that we have explored in this book so far. Building on a decade's

Towards foundational water management

work on the foundational economy – the set of essential infrastructures and systems such as water and sewerage[1] – we call this approach 'foundational water management'. Recognising that this ideal is utopian, we will argue that the foundational question is not how we can completely realise foundational water management, but how it can be approached by a process of adaptive reuse that changes the trajectory and modifies the tendencies of the existing system. Hence the proposal for four enabling technical reforms: the first two are reform of water charging and public ownership as a backstop against the irresponsibility of financialised owners; and then, more ambitiously, the second two are catchment and national planning of the water and sewerage system. As we will see, much of this involves learning and experimentation which covers both the 'what to do' and the equally critical 'how to do' issues.

Where do we want to go?

As liberal collectivists, our foundational aim is the individual freedom of citizens to live lives they have reason to value.[2] A prerequisite for this is collective action to secure the foundational basics necessary for liveability and sustainability, which clearly includes utility services such as water and sewerage.[3] For purposes of political motivation and mobilisation it is important in the case of water (or other basic services) to continue to assert that these are essential so that individuals have a right to these services. Equally it can be argued that water is part of our commons and a resource which should be managed for the long-term public good. This is the case because the built systems of pipes and works which provide daily access to clean water and wastewater treatment are embedded in the water cycle and interconnected with much else, including the ecologies of inland and coastal waters in an age of nature emergency. Therefore, the public interest cannot be defined simply in terms of the quality and cost of water as a utility service. Equally important is adaptation to climate change, including resilience against flooding and mitigation of nature emergency in a world where recreational access to green and blue infrastructure is increasingly valuable for citizens.

Recognising these claims and interdependencies does not resolve the practical issue of how a specific service such as water and sewerage should and could be provided in a particular economy, polity and conjuncture. But we can start with some generalities on foundational water management and argue that this would ideally satisfy five desiderata. Taken together, these provide a benchmark against which we can judge actual provision and set a direction of travel for future provision. The direction of travel would have to include citizen participation in working out what each of these might mean in practice and what some of the trade-offs might be. But the five desiderata are in general terms an uncontroversial specification because we could all agree that the water system should meet the following criteria.

1   *Technically capable*, that is, sustained delivery of clean drinking water and disposal of wastewater in the face of both water shortage and water excess under climate change.
2   *Socially and spatially inclusive*, that is, capable of servicing all households to a high standard in all regions.
3   *Financially accessible*, that is, with costs charged equitably so as not to overburden low-income households or undercharge high-income households.

Towards foundational water management

4   *Environmentally responsible*, that is, with potable water sourced sustainably and waste disposed of in ways that are not environmentally damaging.
5   *Democratic*, that is, with an informed and organised public able to influence decision making and the overall dynamic and tendency of the system.

In our view, a foundational water management system which meets all five of these desirable criteria is an ideal vision which can never be fully realised. This would require a clean break from today's mess as described in the previous five chapters, and a complete transition to another stage or state of water history. The idea of transition is popular because it articulates green ambition. But if the aspiration is laudable, the idea of transition from one state to another rests on a misunderstanding of socio-technical change. A clean break is never going to happen because, in practice, the old survives alongside the new, and the new is imbricated in the old. Hence the foundational concept of change through adaptive reuse, which is borrowed from the French architects Anne Lacaton and Jean-Philippe Vassal[4] and overlaps with David Edgerton's concept of innovation as creative adaptation, not diffusion of novelties.[5] This is not an admission that should discourage radicals but a recognition that empowers radicals, because they can pragmatically pursue change in systems such as housing, energy or water, where we cannot simply 'start again' but have to 'start from here'. From this point of view the first step is to understand how an existing system falls short of the five ideals in a specific time and place, and what its trajectory and tendencies will be without intervention. We can then figure out how we can, though technical design and political leverage, move successively closer to the ideal vision.

Murky water

The English and Welsh water system in the second half of the twentieth century was organised around a sanitation problem, and more or less realised the first three of our five criteria (failing on environmental and democratic desiderata). On the basis of our empirical analysis in Chapters 1.3 and 3.3, in a new conjuncture the problem to be addressed is water management under long-run climate change. The old problem of sanitation (symbolised in our time by storm overflows) is now enclosed in this new larger problem. However, addressing the new problem is complicated because, as we have argued in this book, this is not just a system messed up by financialisation and the failure of political and regulatory control. The system has a business model problem, with revenue constrained by the charging system and, as we shall see in the next section, it is moving towards projectification which will add new problems. Our expectation is that the tendency of the English and Welsh water system will be to become much less foundational and meet fewer of our five criteria. As Chapter 3.3 argued, by the mid-twenty-first century the English and Welsh regions will have serious water excess and deficiency issues. Some regions and localities will be better served than others, and as Chapter 4.2 showed, water will be very expensive for low- and/or middle-income households under the regressive charging system. In effect, on its current trajectory, the water system is moving backwards towards meeting none of our five desiderata, making the UK water system more like those in low- and medium-income countries, with service failure, uneven access and lack of resilience.

If we bracket the question of how we find political leverage for reform, the immediate technical question is what changes in policy and practice are necessary for the English and Welsh water system both to defend its twentieth-century achievements

Towards foundational water management

and to meet the twenty-first-century challenges of water management. In foundational thinking, this depends on the analysis of activity and conjunctural characteristics, which is why this last chapter has been preceded by such a wide-ranging analysis in the first five chapters. We have taken in much more history than would be found in a policy report, so that we understand the novelty of our current conjuncture and can try to avoid the recurrent patterns of behaviour that cause history to rhyme. Foundational thinking is concerned with specifics in historical context, which means that there are no generic foundational policies that apply, regardless of circumstances or conjuncture. For example, in thinking about public ownership, we need to understand the specific objectives to be achieved. The technical points of intervention and relevant levers of change can only be found after empirical analysis of specifics as in the earlier chapters of this book.

Four key reforms

Given current English and Welsh conditions, we propose four technical reforms for building a more foundational water management system in the age of climate change. At this point, issues of implementation and the limits on the doable can no longer be avoided: implementation of policy and practice changes always requires a back office with the direction and capability to deliver on front-office promises.[6] As we will argue in the second half of this chapter, the issue here is not simply one of administrative capability and formal governance structures, but of who are the social actors and what are the power relations that determine outcomes. But at this point, we can focus on back-office capability. The first two necessary reforms (progressive charging and publicly owned companies) would be within existing capabilities if there was a change in political direction in Westminster. The more difficult second two reforms (coordination of multiple social actors and the reinvention of planning) are outside the bounds of what is currently possible because they require a change in the political power configuration to allow organisational experiment, as explained later in this chapter. Any attempt to reorganise technical domains in an ambitious way will encounter political resistance and require struggle. But if we bracket these issues, here are four key technical reforms.

1 A progressive (or flat-rate) charging system based on household income is the essential liberating precondition for a foundational water management system. Progressive charging is a game-changer, socially and economically. Socially, it secures equity through a charging system which avoids overburdening low-income households, while ensuring that high-income households pay their fair share for this valuable service. Economically, progressive charging allows the industry to raise a larger revenue fund to allow much higher levels of capital expenditure, which are, as Chapters 3.1 and 3.2 argue, urgently needed. As long as this revenue constraint persists, extend-and-pretend water company under-investment will continue, while private finance will create more toll points in an increasingly fragmented water system, as we will see in the next section. Easing the revenue constraint also then puts finance in its place as the servant, not the master, of renewal priorities. This is necessary because finance-driven distortion of infrastructure renewal priorities is inevitable if additional debt is used to cover the revenue deficiencies that are a feature of the current charging system.

Towards foundational water management

With household water bills now rising significantly, the regressive impact of the current charging system is only magnified. As we showed in Chapter 4.2, introducing a broader, national social tariff would help the poorest but at the expense of creating new inequities in low- and middle-income groups who do not qualify for support. Flat-rate or progressive charging systems can be a powerful technology as they allow more revenue to be raised, while distributing the cost of the water system and its renewal much more equitably. The introduction of progressive water charging is also important because it connects with broader foundational reform issues, which include addressing regressive charging for electricity and gas, where bills are much larger than for water. The precondition for all this is a shift in political discourse on redistribution from poverty to social justice in water and the other on-market utilities. Such a shift would involve broadening the field of the visible so that redistribution is not simply about the poor and the rich. If we want to fund foundational infrastructure renewal in an equitable way, we must tap the income and wealth of managerial and professional households in the top three deciles (not just a small number of the super-rich).

2 Public ownership of the existing regionally based catchment companies is the essential defensive precondition for a foundational turn in the water management system. It is necessary here to begin by insisting that public ownership is a precondition, but that on its own it does not bring financial and public interest benefits. In 2018 the Labour Party proposed a bond-financed, not-for-profit publicly owned water company. As the Dŵr Cymru case showed in Chapter 2.3 on ownership, there is no guarantee that this kind

Murky water

of entity has lower financing costs which could allow more investment and/or lower household bills. Equally, there is no guarantee that public ownership in itself is enough to make a company democratically accountable and a servant of the public interest. However, after a thirty-year experiment, it is clear that private ownership and management is fundamentally unsuitable for water given its activity characteristics, especially in an age of financialisation. As Chapters 2.1 and 3.1 showed, private ownership cannot deliver 'efficiency improvements' in an asset-heavy, labour-light activity; instead, private ownership instals financialised investors who divert from improving operations for public benefit into financial engineering for private returns. What is more, it prioritises short-term horizons in an activity that needs long-term planning. The environmentally responsible behaviour that figures in our criteria for foundational management outlined above will not be the automatic result of public ownership, but it is difficult to imagine this being possible without it.

Against this background, public ownership is, for hard, unsentimental reasons, a necessary defensive move to protect against financial engineering and short-termism. Water charges are already rising to provide more revenue for investment, and any progressive redesign of charging systems to allow more funds to be raised would make water even more attractive for financial engineers such as Macquarie and other investment funds. Chapter 5.2 showed how Ofwat regulation failed to block financial engineering, and regulatory reform cannot be trusted to do the job in the next twenty years. Public ownership can immunise the water infrastructure against financial engineering, and it need not be expensive. Compensation of private shareholders and bondholders should be based on the fact that, as we argue in the next section of this chapter, the state would be buying

Towards foundational water management

zombie companies in an over-capitalised industry. The industry needs restructuring, which in the private sector would involve writing off the equity and writing down the debt. This should be the basis for compensating water company shareholders and bondholders when they are bought out. If the industry's debt is written down before the industry passes into public ownership, the public company's balance sheet would allow it to issue bonds or take out syndicated loans to finance new major projects. This would also mean that projectification – in many ways the next stage of financial engineering, as we analyse in the following section – is not inevitable.

After these two changes to the charging system and ownership, we come to the more difficult moves which require not better governance but a change in governmentality. In free market ideology, dispersed private decision making in product and financial markets is generally supposed to deliver superior results to any form of planning. But water and the other foundational utilities have to be organised as systems of nodes and connections. In the case of the existing private water companies in the UK, we have river-basin-based organisation with an absence of connection between catchment areas, so that less than 5% of water is currently traded between the existing companies. If we are to cope with the new challenge of water management under climate change, we need national coordination of decision making for major interventions, as well as catchment planning which gives nature-based solutions a chance to compete with civil engineering solutions. The tragedy of utility privatisation was that it destroyed the organisational apparatus and administrative capability for planning, which was embedded in the nationalised utility services such as electricity and gas. After the passage of thirty-five years, there is no institutional memory that can be revived. But we must now, in the spirit of Samuel Beckett, try again in the hope of failing better.

3 Coordination of multiple social actors within each catchment area is a new necessity if large-scale nature-based solutions are to become deliverable. Climate change brings a new kind of water management challenge. The problem of water excess with river and surface water flooding, for example, is clearly beyond the traditional kind of self-contained water company whose business is sanitation and whose responsibilities are embodied in the system of pipes that deliver water and take away waste. Managing water excess in a catchment area involves coordinating multiple social actors, including households and local authorities in urban areas and farmers and landowners in rural areas. This is also within a frame where multiple policy objectives on house building, food production and land use intersect. Matters are complicated because much of the development of nature-based interventions, in urban sponge sites or upland afforestation, will be experimental. In current official documents, the issue of coordinating multiple social actors is discussed using the language of 'partnership' and togetherness. This glosses over the awkward fact that the necessary coordination cannot be achieved voluntarily through goodwill when so many actors have divergent understandings and their own priorities or business model exigencies. The downside of dispersed decision making in foundational systems is a siloed rationality of cost and responsibility passing, where what should be everybody's business is nobody's business.

The challenge of water management is increasingly recognised in progressive official policy documents which give glimpses of a coordinated future. But they tell us nothing about how to get there, because planning of coordinated action is outside the

Towards foundational water management

practice and imagination of current UK governmentality. Coordination on a catchment basis needs a plan developed by some kind of organisational guiding mind, with regulatory powers over land use and reuse, plus the financial capacity to reward actors for behaving responsibly and punish them for behaving irresponsibly. The current post-1979 practice of relying mainly on financial rewards to encourage responsible corporate behaviour is costly (directly or indirectly in terms of forgone revenue) and also simply inadequate. If we do not have directive coordination and input of multiple actors across the catchment, the water companies will insist that their responsibility is sanitation (potable water supply and waste disposal). Hollowed-out and cash-strapped local authorities will plead that they cannot afford the kinds of water retention and slowdown technologies that can make urban areas more sponge-like, while farmers and landowners will find reasons to resist the expense of measures including tree planting and water course adaptation, as well as actions to substantially reduce runoffs. Because appropriate solutions will need to be place-specific, bringing in the local expertise of civil society will be an important part of the new forms of coordination and decision making. And including active citizens in this process of coordination is a necessary precondition for our final criterion of democratic water management, as we explore in more detail in section 4 of this chapter.

4 Reinvention of national planning is the second necessity so that we have a cross-catchment level of planning. The current regional catchment organisation of the water companies is not an evidence-based choice but a path-dependent historical accident. It is a stopped-clock legacy of 1970s planning, which created regional water authorities as a first step beyond municipal control and towards river basin management of sanitation. As we saw in Chapters 1.3 and 3.3, catchment areas have very different topographies and variable amounts of annual rainfall. So, water management under climate change requires the coordination of projects and interventions – traditional and nature-based – across as well as within catchment areas. National planning, which integrates intra-regional increase in capacity with inter-regional water transfer projects, is absolutely necessary for dealing with water shortage in south-east and eastern England and ensuring spatial as well as social inclusion. And national planning needs to be as much about financial frameworks as about physical projects. The task of coordination includes setting water transfer prices for inter-regional transfer and supervising a revenue-sharing scheme, if charging by household income involves financial transfers between rich and poor regions. As we saw earlier in the chapter, the existing water companies have long resisted a national social tariff scheme because that would involve financial transfers, given the uneven regional distribution of qualifying households.

While the water management ideal is democratic and requires citizen participation, it cannot be locally or even regionally bottom-up because major interventions need to be nationally planned in the public interest. Regional autarchy will inevitably break down in the next phase as water shortage in the east and south-east becomes more acute. But the outcome under current arrangements will most likely be national mess. Advisory bodies such as the National Infrastructure Commission are not established or resourced for the planning task of managing coherent provision. The Commission has a civil engineering mentality with a bias towards steel and concrete projects. It does not have the power to say 'yes' or 'no' to specific projects,

Towards foundational water management

nor to propose alternatives when it comes to key decisions on issues such as water transfer between regions, new reservoir building and less abstraction in regions. As we argue in the next section of this chapter, project proposals will be driven by water companies with wrecked balance sheets who want to shuffle major investment projects off on to financial consortia. These consortia have the wrong kind of long-term mentality because they create investment projects that function as toll booths, with a multi-year lien on household income, and share a bias towards steel and concrete (not nature-based solutions). Financial products may work reasonably in funding the provision of some consumer goods and services, but if finance interests dominate and substitute for planning of the water system, the results will be completely dysfunctional.

Having set out these four essential moves, many issues remain unresolved. Most obviously, how should the division between catchment and national planning responsibilities be organised in terms of corporations and functions? It probably makes sense to think of some kind of high-level authority developing integrative national plans, with lower-level bodies responsible for developing and delivering regional plans. But there are many choices within this broad division of responsibilities. Or again, how to deal with the cost inflation inherent in the current British system of multi-level sub-contracting in civil engineering projects based on chargeable hours and materials on a 'cost-plus' basis? Should the British create a national corps of engineers to design major renewal projects and then supervise major contractors, much as Bazalgette did? This is the route that the Italians have adopted for the rebuilding of their motorway system after the Morandi bridge collapse. There is much to be said for this kind of innovation, which would be as much about rebuilding lost expertise as about cost management in renewal of the water system.

There are, then, all the issues around the reinvention of regulation. In foundational thinking about the renewal of systems, the primary emphasis has to be on planning as the coordination of action, not regulation as the policing of behaviour. But it should not be assumed that removing the profit motive guarantees responsible and ethical behaviour on the part of public companies or any other kind of organisation, especially after things have gone wrong and there is blame to be apportioned. The independent regulatory policing of behaviour is always necessary. However, the obsolete Ofwat brief of balancing investor returns and consumer prices should be buried, while the problems raised by the under-resourced Environment Agency need to be addressed. After a reform of household charging which frees up the revenue line, we need a return to something like the EU directives which, in the 1990s, mandated compliance with technical standards (adjusted for variable regional circumstances and with latitude for how those standards were to be reached). Regional plans should include delivery milestones with exacting accountability, and there should be regulatory power to call in and question company plans in the long term for at least twenty years, not for five years as under the existing PR system.

6.2 Zombie companies and projectification

There is a painful contrast between the right moves outlined above, which would take us closer to the foundational vision, and what we are about to receive in the 2020s. As long as Defra and the UK mainstream political classes will not contemplate public ownership, the privatised companies will continue as debt-burdened zombies whose balance sheets cannot finance major projects. The logic is what we call *projectification*, with one-off private vehicles used to finance projects, as we have already seen with Thames Tideway. This 'super sewer' project adds surcharges to customer bills which, in the absence of progressive household charging, will have to be managed by further extension of social tariff schemes, with all the limits we noted in Chapter 4.2. Projectification used on a much broader basis will move us further away from the vision of foundational water management while falsely appearing to deliver progress, with press releases and speeches celebrating how we are now building water transfer schemes and reservoirs.[7]

Zombie water companies

The idea of a zombie company is well established in popular investor discourse, where it denotes an individual company that cannot invest or grow because it earns only just enough to cover operating expenses and service its debt.[8] Corporate zombyism has two preconditions – not enough cash from operations or too much debt on the balance sheet – which are interconnected but can be separated analytically.

1 A company that generates limited cash from operations can cover this year's operating expenses but cannot fund the capital expenditure necessary to stay in business. Thus, small car companies such as Austin Rover can survive by running existing models down their lines. But this does not generate

Murky water

    the cash to cover new model development costs in an industry where models have to be replaced every five to seven years.

2 A company that is over-capitalised is burdened by financing costs, which usually means excess debt, interest payments and a wrecked balance sheet. This leads to near-death crises regarding credit ratings and refinancing, without any capacity to pay off the debt or execute any strategy except cost reduction. Using the criterion that the company has generated insufficient cash over the last three years to cover its interest obligations, the zombies currently include Manchester United and Telecom Italia.[9]

If we leave aside the technicalities, it is clear that Thames Water and Southern Water are both completely zombified. In an industry where water company credit ratings have deteriorated over time, both Thames[10] and Southern Water[11] bonds have junk status. Thames Water has been struggling to make payments on £19 billion of debt since its holding company failed to pay the coupon on a £400 million bond in 2023.[12] In early 2025, when it was expected to run out of cash in less than two months, Thames went to court to secure its right to a £3 billion loan from its class A bondholders, who were prepared to lend at a hefty 9.75% plus fees, while its class B bonds with less security were trading at less than 20 pence in the £1.[13] Southern Water has attracted less attention because it is not one step away from liquidation. But on the same day that Thames got lower court approval of its £3 billion loan, Southern Water announced that its fund investor owners, led by Macquarie, which formerly owned Thames Water, would inject another £900 million of equity.[14] The extra debt for Thames or equity for Southern

Towards foundational water management

keeps things going but adds to long-term problems by increasing the stock of finance capital which has to be serviced.

More broadly, the problem in water is that we have not just zombie companies but a zombie industry, given the inability to generate enough cash from operations to cover long-term investment. As readers will recall from Chapters 2.1 and 2.2, the underlying business model problem of the industry is that water is a physical investment-hungry activity. Under the existing revenue constraint, it cannot meet both the requirements of capital expenditure on physical investment and the service costs (in interest and dividends) of finance capital from its operating cashflow. Over the period 1989–2023, capital expenditure amounted to 78% of cash from operations, with finance capital costs amounting to a further 63% (split 29% on interest and 34% on dividends). If we add up all of these calls on water company cashflows from operations over the period 1989–2023, we find that they amount to 140% of the cash generated from selling water services, thus creating a large deficiency. This gap was managed in two ways: by rationing physical investment (which prevents the funding deficiency from being even larger), and by taking out £61.8 billion of debt.

As Chapter 2.3 showed, this problem of balancing physical investment needs and financial capital claims would have existed with or without the profit motive and financial extraction, but financialisation certainly increased borrowing levels and ran up the debt with little thought for tomorrow. As a result, the industry has ended up grossly over-capitalised with a pile of debt it cannot easily service from revenue. The debt outstanding per £1 of sales revenue increased from 40 pence per £1 in 1991 to £3.60 in 2010, and then to roughly £5 of debt for every £1

Murky water

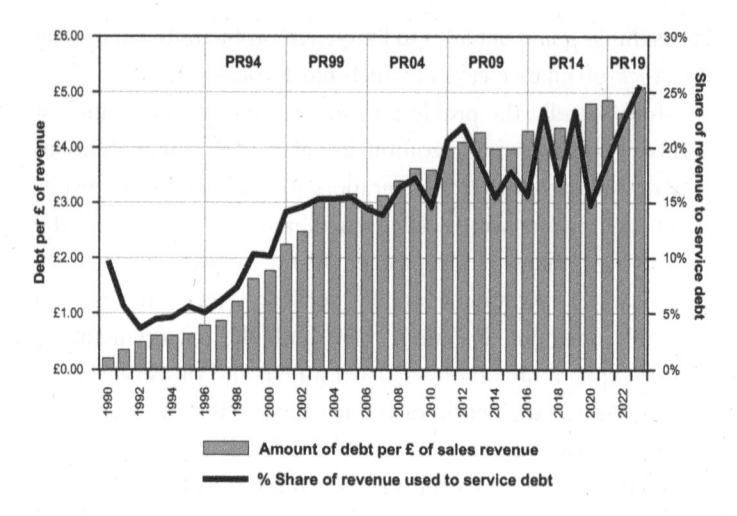

Exhibit 6.1 English and Welsh water companies' debt and debt servicing, 1990–2022[15]

of sales revenue earned by 2023, as shown in Exhibit 6.1. Even with low interest rates in the 2010s, debt service charges in the form of interest payments inevitably absorbed an increasing share of sales revenue, with interest claiming 14.7% of sales revenue in 2010 and 25.8% of sales revenue in 2023.

It is not simply that the water companies have a lot of debt which is not being repaid. The debt mountain will substantially increase over PR24 up to 2030, as the water companies plan to take out more new debt. The plans for increased customer charges and some restraint on dividend payments will help, but physical investment will also increase in PR24, so the basic problem of not receiving enough cash from operations will persist. On the original company submissions to the PR24 process, the planned physical investment, interest charges and dividends added up to

Towards foundational water management

169% of projected cashflow from operations from 2025 to 2030. This has to be covered by taking out more debt. In their PR24 plans, the ten water companies proposed to raise £25.3 billion of new debt between 2025 and 2030, some 66% more than the £15.2 billion of new debt raised in PR19 between 2020 and 2025. After various adjustments, Ofwat's final determination knocked back the increase in new debt required in PR24, so that it was to be about 60% above the PR19 level.[16] Even so, the additional borrowing in PR24 up to March 2030 is expected to increase the stock of outstanding debt for the ten water companies from roughly £70 billion to roughly £95 billion.[17]

In a complicated world, the new debt requirement may be higher if interest rates rise and price inflation takes off again. These are real possibilities given geopolitical uncertainties and the ever-present danger of unexpected shocks. Only around 10% of water company debt is at floating rates, providing short-term relief against rate rises, but around one-quarter of the total borrowing – some £17 billion – needs to be refinanced in the next five years at whatever interest rates are prevailing.[18] The near-certainty is that prevailing interest rates in the later 2020s will be higher than they were in the 2010s. The other complication is inflation. Currently 51% of outstanding debt as at 31 March 2024[19] is linked to the retail price index and repayments can rise or fall with inflation rates. In 2025 an inflation shock has passed (or at least subsided), and most companies can refinance existing debt and raise new debt without paying punitive rates. So, the industry as a whole is not zombified to the point of struggle with ongoing cash crises and junk bond ratings, as with Thames and Southern. But the industry's financial position is increasingly precarious as the debt mountain will certainly increase and the interest charges could escalate in line with geopolitical shocks. All of this means that the industry is increasingly unsustainable in financial terms, and there is no planned event or process that will allow a reset.

Zombies figure in our broader culture as much as in investor discourse. The zombie in movies or the revenant in folklore dramatises the idea of the living dead in a way that adds another dimension to our understanding. In popular culture, the zombie is an animated corpse that lacks all the ordinary qualities of agency in judgement and restraint that sustain our social lives. The zombie is not only humanly incapacitated but is also a violent and dangerous social disrupter. The wisdom of the ages in European folklore and ancient burial practice is that it would be best if the dead departed and could not rise again to have an afterlife in a half-animated state, behaving in ways that are a threat to the rest of us. This is certainly true of the zombie water industry. Half-dead companies with wrecked balance sheets cannot borrow to fund the necessary investment, meaning that investment has to be carried out one large project at a time through private finance vehicles. The disruptive consequence is projectification, which undermines the coordination and improvement of any kind of system and has costs which are political, social and economic.

Projectification or more financial engineering

When Ofwat's final PR24 determination was announced in late 2024, the media announced that the average customer bill would increase by 36% by 2030.[20] With inflation adjustment and front loading of PR24 increases, in early 2025 Thames Water customers were shocked to receive bills which year-on-year were nearly 50% higher.[21] They would also continue to pay surcharges for the Thames Tideway super sewer after it became fully operational in 2025 for more than a hundred years. They had since 2016 already paid surcharges of £160 per household customer towards its construction costs.[22] That was all the information that was in the public domain by early 2025.

But effectively hidden away in the thousands of pages of PR24 documentation, which almost nobody reads, was a list of 27 privately financed major projects undertaken for English water and wastewater companies for reservoirs, water transfer and sewerage treatment works.[23] These are important because they will lead to extra charges on bills which do not figure in the PR24 settlement. The development costs of these private finance projects are being paid by the water companies and charged to customers in the bills agreed under PR24, but everything else by way of capital, operating and maintenance costs will be recovered by surcharges on water company bills in future years, over a term of 25 to 125 years. The total expenditure for construction, operation and maintenance of these private finance projects is officially estimated at £47.9 billion, of which just over £1 billion is for in-house projects.[24] The whole-life capital expenditure of these 27 projects is £22.4 billion, which is equivalent to two-thirds of the declared total PR24 enhancement expenditure of £34 billion on upgrade projects.[25] And £14.5 billion of the privately financed capital expenditure will start in the PR24 period by 2030. The Cunliffe Commission Call for Evidence notes the existence of these schemes,[26] and Dieter Helm questioned their rationale in a 2024 blog,[27] but their scale and implications have generally remained hidden from view and from debate. All the major projects agreed by late 2024 are shown in Exhibit 6.2, with a breakdown between three different types, explained below.

Murky water

	Project Description	Whole life Totex £m	Capital Expenditure £m
Anglian	Reservoir	4,071	3,271
Anglian	Reservoir	4,158	3,766
Thames	Reservoir	7,523	2,662
	SIPR projects: sub total	**15,752**	**9,700**
Affinity and Severn Trent	Water transfer	1,540	467
Affinity and Severn Trent	Water recycling	640	176
Anglian	Desalination	2,337	727
Anglian	Desalination	2,205	686
Northumbrian	Water transfer	389	174
Severn Trent	Reservoir	812	62
Severn Trent and Canal Trust	Reservoir	419	309
Severn Trent and Yorkshire	Source and transfer	491	316
Severn Trent and Yorkshire	Reservoir	738	290
Southern	Water treatment/water transfer	755	341
Southern	Water recycling	1,295	314
Southern	Transfer and water recycling	3,031	1,236
Thames	Water transfer	2,188	1,006
Thames	Water recycling	3,511	994
Thames	Water transfer	1,729	635
United Utilities	Water transfer	4,310	1,537
Wessex and South West	Water transfer	353	323
Wessex and South West	Reservoir	1,728	1,218
Wessex and South West	Reservoir/water transfer	839	680
Yorkshire	Transfer and water recycling	306	163
Yorkshire and Northumbria	Water trearment	310	220
Yorkshire and Northumbria	Water transfer	1,146	384
	DPC projects: sub total	**31,072**	**12,257**
Severn Trent	Water treatment/water transfer	145	124
Thames	Water recycling	989	301
	In-house: sub total	**1,134**	**425**
Grand total		**47,958**	**22,382**

Exhibit 6.2 Strategic Infrastructure Procurement Route, Direct Procurement for Customers and in-house major projects agreed by Ofwat (in 2022/23 prices)[28]

Towards foundational water management

Where do these projects come from? The three water regulators – Ofwat, the Environment Agency and the Drinking Water Inspectorate – together review the companies' Water Resource Management Plans which establish an urgent 'needs case' for major infrastructure projects, defined as those with a total expenditure over £200 million.[29] Water companies then suggest specific projects, which they are generally unable to fund 'in-house' by raising substantially more new debt because they are already heavily indebted and committed to catch-up investment under PR24. Hence the funding gap is filled using one of two models, both of which involve establishing standalone project companies that can raise debt. Specified Infrastructure Project Regulations (SIPR) schemes were originally introduced in 2013, and Direct Procurement for Customers (DPC) schemes were introduced a few years later as part of PR19. Under both models, private financial consortia design, finance, build, own, operate and maintain major infrastructure projects. The incentive for private finance is a guaranteed, inflation-proofed income from a surcharge on customer bills. The SIPR is a licence scheme for the largest schemes – such as major reservoirs projected to cost several billion pounds – which set up the licence-holding consortium as an 'independent regulated water company'[30] with cost recovery by surcharge for a very long period: in the case of Thames Tideway this is 125 years. DPCs operate with a contractual arrangement rather like the PFI contracts that were finally discontinued in 2018 after accumulating criticism and a decisive National Audit Office report on the dismal results of both the original PFI schemes and their successors under PF2.[31] This did not discourage the water regulators and the companies who, as with PFI, are offering contracts which will typically run for twenty-five years.

Murky water

Chapter 3.2 described Water UK's investment plan as 'extend-and-pretend'. We can now see that the industry's ultimate pretence is that the water companies are the agents fixing the industry's under-investment problems. Under PR24, the reality is that private finance consortia will do more of the heavy lifting under Ofwat's guidance, which stated that 'DPC will apply by default for all discrete projects above a size threshold of £200m whole life totex'.[32] A total expenditure of £200 million over a project's economic life is, of course, a low threshold in this asset-intensive activity. As Exhibit 6.2 shows, many projects which are much smaller than multibillion pound reservoir schemes will now be DPC-funded because the water companies are sinking under the weight of their accumulating debt. The process of projectification applies more financial engineering as the fix for the zombie company problems that financial engineering has already created. Here are six reasons why these private finance deals are not in the public interest:

1 *Undemocratic in conception.* The DPC and SIPR arrangements were devised as a fix by the water companies, regulators and Defra without any public announcement which would allow broad-based political deliberation and decision. The urgent and immediate issue is whether a large-scale extension of private project finance using SIPR, DPC or variants thereof is in the public interest, given earlier bad experiences with PFI. This whole process is driven by debt-burdened zombie companies that are unable to put major projects on to their wrecked balance sheets. The underlying issue is whether the publicly canvassed alternatives to PFI should explicitly include restructuring the water companies with debt write-downs and/or taking them into public ownership.

Towards foundational water management

2 *System-disrupting and unplanned.* Water management in every catchment requires an overview of necessary provision, leading to a catchment plan with a considered bundle of steel-and-concrete and nature-based solutions to be progressed in a coordinated way. The water companies and private finance are instead unbundling, as water companies make one-off proposals for fixes at the beginning or end point of their pipe networks. The projects for reservoirs and sewerage treatment and water transfer projects neglect everything that comes in between, before or after their networks. These fixes are hugely biased towards steel-and-concrete construction because that is what can be packaged into discrete fundable projects. Too many new sewage holding tanks, and not enough sponge or water-retention measures, takes us further away from the planned and coordinated system we want.

3 *Disadvantage by contract terms.* The terms of any licence or contract reflect the balance of power and knowledge between the two parties. In this case, there is an asymmetry of power and knowledge which benefits the private finance consortium in any project. As long as private restructuring or public ownership is ruled out, if the backlog of construction is to be fixed, the water companies and Defra have no alternative but to do private finance deals on major projects. This power disadvantage is reinforced by knowledge when it comes to licence or contract terms. The advisers of private finance will be well along the learning curve after working on a succession of broadly similar contracts, and there is no evidence that regulators or individual companies can mobilise countervailing expertise. Earlier use of PFI in schools, hospitals and elsewhere provides many examples of contracts disadvantaging the public.

Murky water

4 *Unaccountable in operation.* Delinquent privatised water companies are accountable to some extent because they are, under Parliament, subject to regulatory fiat. In the regulated system, new offences can be added, sanctions such as fines can be varied and companies operate under licences whose withdrawal would effectively close the company. SIPR in the largest projects creates licence holders such as BTL in Thames Tideway who are classified as 'independent' regulated water companies. But DPC projects – which constitute the majority of those listed in Exhibit 6.2 – are outside the existing regulatory framework. Here, the only safeguards are gateway review by Ofwat and competitive tendering by different consortia, in a set-up that looks rather like defence contracting and is no guarantee of good outcomes. Even if DPC is brought into the regulatory frame, individual projects will be governed by the terms and conditions of the original contract. So the courts are the arbiter in DPC, and the public interest can only be expensively (and narrowly) litigated.

5 *Inflexible as needs change.* In the long term, water management needs will change and so the water managers in any catchment will need to vary their use of project facilities. Buying out the financiers holding the contract will ordinarily be prohibitively expensive because the projects are viable businesses with a guaranteed stream of income. As in existing PFI deals, the financiers will be happy to oblige with contract variation, provided they are paid handsomely and receive compensation for any costs incurred or revenue lost. The fundamental problem here is that, under private finance projects, ownership is transferred to the finance consortium for the term of the contract or licence, and the water

Towards foundational water management

company loses the right to do what it wants with the asset. In this respect, the DPC and SIPR forms of private finance are a poor substitute for a syndicated loan where (as in a domestic mortgage) the lender holds security and assumes ownership only in the event of default on payment.

6 *Toll the paying customer.* The customer pays the costs of construction, financing, maintenance and operation through a surcharge on the bill for the term of the contract which, as we have seen, will last between 25 and 125 years. SIPR and DPC projects effectively create a series of toll points where the customer pays private finance on a cost-plus basis. All the original privatisation promises of efficiency gains and mechanisms for enforcing cost reduction are jettisoned. From previous private finance experience, the asset may be handed back in poor condition because there is little incentive to spend on maintenance as financiers approach the end of their term of beneficial ownership. Even though customers will have paid directly over decades through surcharges, the asset does not pass into public ownership but will be put on to the balance sheet of the water company, which has done nothing except act as an intermediary without taking any risk.

6.3 Power configurations and policy pathways

The analysis in this book suggests that British governments are on a trajectory where they will do lots of the wrong things in water management, such as through projectification and social tariffs, without introducing progressive charging plus public ownership. These choices will move us further away from foundational water management. Nor can we have confidence that existing electoral processes are sufficient to produce a majority government or coalition in Westminster that will act to realise more foundational water management. The roadblock can be characterised as something like oligarchy, which takes the form of an elite power configuration that delivers what business and financial elites want. While members of the public protest, there is no formal provision for intervention by active citizens. Thus, as we observed in the previous section, projectification will send us further down the wrong policy pathways, even as policy elites assure us that the problems are being tackled. This section unpacks and explains our claims about power configurations and policy trajectory, which is the broad political context that explains current water policy and practice.

If democracy is the rule of the people, practically it means government by representatives of the citizen body, which tends to be what Benjamin Barber classically termed a 'thin form of representative government'.[33] In this case, the power of political elites is subject to electoral consent but there is very limited participation by active citizens. The logical corollary of limited citizen participation is the exercise of power by elected representatives of the citizens acting in political assemblies, and more generally by unelected elites who occupy institutional positions of power in the machinery of government, business and finance. The outcome is that the collective provision of foundational goods and services is done to and for citizens, not with citizens.

This outcome is seldom the result of a considered delegation of power by citizens to parliamentarians and experts, but it is a default in large, complex societies. The structural enablers of this governance operate both bottom-up and top-down.

Towards foundational water management

Bottom-up, most UK citizens are preoccupied with work and family life so that, as Mick Moran observed, time-consuming politics 'is a minority interest' for maybe 100,000 party and NGO activists out of a population of more than 60 million.[34] Top-down, there is always professionalisation in large political and economic organisations, and bureaucratic authority is hierarchical in ways that give considerable power to those at the apex of organisations. Hence the space of participation through deliberation and decision making is by default occupied by policy elites, not active citizens. Elections mostly offer a choice between party variants on the same centrist agenda, though they are sometimes an opportunity for populist parties to mobilise consent around the idea of 'throwing the scoundrels out', but without much control over what comes afterwards.

Foundational history is, then, the issue-based, bottom-up struggle of citizens to manage and direct control by policy elites. This struggle always has its heroes who stand up to power by organising resistance. In our own time the UK has a roll of honour stretching from Pwyllgor Amddiffyn Capel Celyn, the villagers' defence committee which unsuccessfully resisted the construction of the Tryweryn dam in the 1960s,[35] to Save Windermere, the grassroots organisation which in 2025 won 'first step' improvements on ending gross sewage pollution in Lake Windermere.[36] Win or lose, organised resistance not only changes how people think but can also change the rules of the game, as it did in Wales. But on water and sewage, the deep and complex problems are such that there need to be many (and coordinated) wins, and we can only begin to do this if we understand the two key variables that jointly determine outcomes:

Murky water

1. *The power configuration*, which is the relations of hierarchy, competition and cooperation between different elite groups operating in the space of deliberation and decision making. Large, complex societies do not have one unitary power elite but several distinct elites – political, technical and financial – with different knowledge and institutional power bases. Policy outcomes thus partially depend on the internal power relations between these elite groups. These vary according to time and place in ways that directly set the agenda for what gets done or not done.
2. *The presence or absence of organised bottom-up political pressure from citizens*, which when strong can influence the agenda of elite policymakers. Such organised pressure can be exerted in many different forms. If political parties are organised democratically, the pressure can come from party branches and rank and file members. Equally, the pressure can come from trade unions in an organised labour movement, or from civil society pressure groups. The extent and nature of this bottom-up pressure on policy elites varies greatly according to time and place, with implications for the forms of foundational provision and the priority of foundational improvement.

On these assumptions, foundational provision mutates from one conjuncture to the next as the power relations between different elite groups shift, as does the extent of organised accountability to the masses. This is important because the policy pathway in each conjuncture is determined by how this play of forces within the political system is resolved differently over time. The agenda and policy pathway are determined by the internal balance of elite forces inside the power configuration and the nature of

Towards foundational water management

citizen pressure from outside the power configuration. This allows us to understand the general contrast between the 1945–75 period of heroic infrastructure construction, which delivered a new foundational settlement that benefited the mass of citizenry, and the post-1980 period when the little that got built was (and is) whatever suits private business and finance interests.

The thirty years of foundational building after 1945 were underpinned by a power configuration of functional intra-elite relations between the dominant political and technocratic elites. This was subject to national external pressure from organised labour but lacked local democratic input or scope to shape top-down power. In this configuration, the scale of central and local state ambition and the capacity for delivery are memorialised by the 32 new towns built on greenfield sites from Stevenage to Milton Keynes which had an initial target population of 250,000. After 1945 the country was connected by new national systems such as motorways and an electricity grid from large coal-powered power stations, while large social housing developments and NHS regional hospitals were distributed across the country. Under the aegis of the local state, as we outlined in Chapter 1.2, much of the water infrastructure that we use today was built during this period of heroic construction. The point is illustrated in Exhibit 6.3, which shows the age of UK reservoirs, and demonstrates that most of the capacity dates from the second half of the twentieth century.

In this period of foundational building, the government machine had both the administrative capability and adequate taxpayer funding to turn modernist blueprints into structures on the ground, and politicians licensed technocrats to design and build new systems. External agenda-setting pressure came through the Labour Party and the trade union movement, though the

Murky water

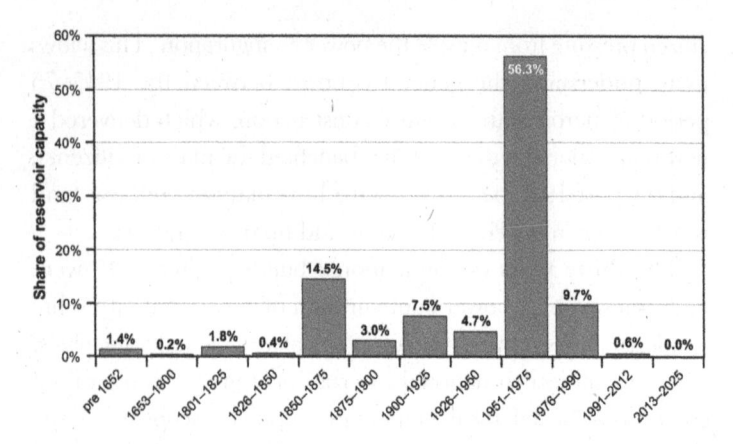

Exhibit 6.3 Vintage of reservoirs in England and Wales, showing the percentage of the 2025 reservoir capacity added in each time period, 1652–2025[37]

political classes largely monopolised deliberation and decision making because provision was being done for and to citizens. For example, top-down housing policy overrode local objections from the Stevenage Residents Protection Association,[38] just as it ignored tenant needs in Manchester's Hulme Crescents social housing development, which was deemed unsuitable for families within two years of completion.[39] And, as we saw in Chapter 1.1, reservoirs involved land clearances from the 1880s to the 1960s.

After 1980 construction slowed down in a new, foundationally dysfunctional power configuration as increasingly dominant finance interests determined what got built. Political and technocratic elites were demoted in a rearrangement of elite hierarchy, in which increasingly dominant private finance promised to fix the problems that arose from the limitations of state revenues. At the same time, external pressure was removed with the decline of the organised working class as a major political actor. Experts

Towards foundational water management

of all kinds became dependents as they worked on problems that politicians defined and/or on contracts from private finance, as with the DPC water projects described in the previous section. Within this new, post-1980 framework, we can explain why the UK seems unable to sort out the problems of the water industry or, more generally, act for climate change adaptation. What we have in general – and specifically in water – is a power configuration characterised by a dysfunctional relation between expert, political and financial elites (under the tutelage of finance), with the citizenry involved in only a limited way. The elite dysfunction is three-dimensional:

1 UK politicians have recognised the problems of climate change though virtue signalling targets for medium-term transition to net zero. These create increasing political difficulties regarding targets that have not been met, amid resistance to the costs and claimed inconvenience of net zero. With mitigation failing, we now have accumulating adaptation problems in water and many other areas. But, as noted in Chapter 5.1, Westminster politicians are mainly in issue management mode in responding to public disgust about storm overflows. Official reports increasingly recognise water management issues, especially water shortage as a constraint on construction. However, there is limited focus on how to tackle water excess – especially surface flooding – because that would require planning and coordinated action.
2 Experts and technocrats are commissioned by politicians to produce reports which format the problem and make recommendations that are politically acceptable as fixes for a manageable issue. This is explicitly so in the Cunliffe

Murky water

Commission's brief, or implicitly so in Defra and NIC reports. When recommendations do not turn into action, experts on the sidelines act as a kind of Greek chorus lamenting inaction and threatening punishment, as with the Office for Environmental Protection. So, technocracy is turned upside down because experts are not leading players but an adjunct to political calculation; and when they get ahead of policy or practice it is generally only in an indicative way.

3. Fund investors in equity and bondholders are now the dominant elite group. As we saw in the previous section, when the state can't pay and won't pay, consortia that mobilise private capital determine what gets built and how, in a process of projectification. Finance will build whatever is fundable on a project-by-project basis, which of course undermines coherent system renewal. Projectification brings private capital, with bond finance allowing build-now, pay-later projects, plus private ownership with shareholders bringing financial engineering and creating toll points. After Thames Tideway, DPC and SIPR projects are the new frontier of opportunity for private capital in water. As with PFI hospitals and schools, the public downsides will become clear over time.

The downscaling of technocratic ambition in the twenty-first century is incontrovertible and the political behaviours which demonstrate the dominance of finance in the current power configuration are ubiquitous. In many policy areas we see the same retreat from large-scale construction. For example, the Labour government in 2024 reprised the idea of new towns, but now the promise was 'up to 12 new towns', each with 'the potential for at least 10,000 homes' and no guarantees on homes

Towards foundational water management

for social rent.[40] In water we have construction stop-and-go at the behest of finance. It is not just that, as Exhibit 6.3 shows, less than 1% of the current reservoir capacity was added after 1990. Now in the 2020s, as the previous section demonstrated, stalled construction is restarted via projectification because that brings new profit opportunities for finance consortia. Moreover, discussions in 2025 around how to recapitalise Thames Water, including a bid from private equity fund KKR, show the power configuration in action. How else can we explain the way in which a finance market-led restructuring was preferred as an alternative to credible and not very radical options such as putting the insolvent Thames Water into special administration (with or without temporary public ownership)? The language of investability and exaggerations about the risk and cost of public ownership provide an all-purpose justification for the subordination of political judgement to the needs of finance.

Our high-level account of post-1980 developments must be an interpretative sketch, and we cannot provide the detailed, evidenced analysis that sustains the rest of this book. However, our account of the power configuration is coherent with other threads of our argument, as, for example, in the appendix to Chapter 5 on abstraction, with Defra and Ofwat desperately trying to avoid specific programmes of measures for water courses. And we would note that our argument in this section about power does connect with insights in recent academic research and journalism. Circulation of personnel between elite roles and groups has become an important mechanism whose effect is to homogenise judgement and create a shared world taken for granted by those who pursue centrist careers in Westminster and Whitehall. They can promise 'change' but will mostly deliver more of the same. In research, a key text is Aeron Davis's book on the Treasury since 1976, which describes a process of elite capture, as the 'Treasury brain' favoured the finance sector and its private capital solutions, while Treasury ministers increasingly came from finance sector professional backgrounds.[41] In journalism, there is a stream of stories about the rise of professional lobbyists whose main intermediary role is to rationalise the needs of corporate business and finance to politicians. These intermediaries are increasingly important as candidates and sitting MPs in all political parties. For example, in the 2024 General Election, 103 lobbyists stood as candidates, and a professional lobbyist was 27 times more likely than a teacher to stand as a parliamentary candidate.[42]

Within this new elite power configuration, the empirics show that citizens are marginalised as worried and confused observers on the sidelines. In 2024, according to More in Common polling, 65% of respondents were worried or somewhat worried by climate change. Less than 20% rated climate change as a top three issue of concern, and other issues such as the NHS and immigration were more important as drivers of voting behaviour in the 2024 General Election. More in Common's explanation is that, in 2024, 'cost of living is the key lens through which people see climate action' and support for green policies is partly based on hopes of lower energy costs;[43] they did not add that the primacy of the cost of living issue reflects a political failure to control the cost of essentials such as housing and energy. Less responsible politicians then try to mobilise citizens by playing on sentiment about net zero and the (avoidable) costs of mitigation, so that the unavoidable costs of adaptation as in the water and sewerage system are not a public issue.

Towards foundational water management

If we want to move towards the vision of foundational management set out in the first part of this chapter, we need social actors who can organise and provide external pressure to reset the agenda and deliver change. This is especially relevant in water, where an increasing number of civil society groups are already agitating for change, and Westminster politics will increasingly have to try to contain and manage these groups. The next section considers the achievement of civil society groups and the disruptive potential of a social movement for water.

6.4 Towards a social movement that challenges power

The obstacle to progress in water is what UK financial, technocratic and political elites can all too readily agree on. The Cunliffe Commission's report that accompanied the call for evidence in early 2025 provides an updated restatement of that consensus. Engagement on all manner of issues is encouraged but, at the same time, the priority of keeping the industry 'investable' means that politicians will sanction very little unless it suits finance. The dominance of finance is accompanied by the absence of an organised and disruptively focused citizens' opposition. Given this starting point, moving towards foundational management of water requires the challenging, not the conciliation, of power. The question then is how can we disrupt the power configuration outlined in the previous section so that that our foundational proposals for reform of the water management system become progressively more realisable? Radical reform requires not just decisions but spaces of deliberation where citizens can form and express their views without manipulation, and that has to be part of a larger project of obtaining political leverage.

On the left, this is conventionally thought of as a problem of representation mechanisms and institutions that would remedy a democratic deficiency. As the campaign group Compass argues: 'the group of people with a seat at the table in the governance of English water does not include anyone with a legitimate democratic mandate to serve our people and our environment'.[44] In a similar vein, the 2018 Labour Party public ownership proposal was to replace the private water companies with ten water authorities governed by tripartite boards of councillors, trade unionists and NGO representatives. These boards would be advised by 'water observatories' for a democratic debate about water issues. In deepening crisis, Compass in 2025 launched a shadow board for Thames Water,[45] on similar lines to the 2018 proposal, and the Labour MP Clive Lewis in March 2025 sponsored a private water bill which calls for a citizens' assembly on water ownership.

These ambitions for economic democracy, which is 'a critical cornerstone (and pre-requisite) of genuine political democracy',[46] are important interventions. But at the same time, we can question whether the new seat at the governance table approach is now adequate, given the problems we have explored in this book, especially around household charging and the larger challenge of planning for water management under climate change. The focus on governance reform reflects both the misconception that the basic problem is sanitation, and the false assumption that not-for-profit, publicly owned water companies would have a sustainable business model. In the UK's highly centralised polity, a regional water authority might be able to determine bill levels, but it would not have the authority to make the essential changes to how households are charged and introduce a more progressive charging system. Equally, a publicly owned regional water company would not (and almost certainly should not) have planning powers over landowners and local authorities to deal with catchment water management. That would require a new kind of organisation with a broad overview, not an infrastructure operating organisation with a new, more democratic board.

Citizens' assemblies are distinctive because they effectively create a new table with citizen seats. There is strong evidence that these assemblies can produce informed citizens who make intelligent majority recommendations after hearing expert evidence.[47] The problem is with their ability to shape political decision making. For example, the Climate Assembly UK in 2020 produced an environmentally sensible and politically saleable set of recommendations across eight broad areas, based on six weekends of hearing evidence, inquiry and deliberation. For example, the Assembly agreed that flying should continue but become more expensive and that frequent flyers should pay more, while the increase in passenger numbers should be related to technical progress in reducing emissions.[48] This citizens' assembly had promising beginnings given that it was commissioned by six Parliamentary Select Committee chairs, but Westminster politicians still ignored its recommendations.[49] Thus, on the issue of flying, the Labour government was by 2025 promoting new runways for airport expansion. There are examples of where citizens' assembly recommendations do result in action, but this is generally only when it suits the political classes to delegate decision making which would otherwise split them, as was classically the case with abortion reform in Ireland.[50]

Changing accountability arrangements in some organisations and new kinds of representation will deliver little unless the change is animated by a broad-based and disruptive social movement. Such a movement has to be sustained over a long period and explicitly include the invention of new institutions and the reordering of the agenda at all levels of government. On this basis, how can thinking in common inform and empower actors who are on the margins but could organise to create a social movement around water management; and how can this then disrupt the power configuration and open up the political space for new policy pathways in water management? This is an ambitious challenge, but there are already promising developments such as the London March for Clean Water in November 2024, which brought together more than a hundred organisations,[51] or the End Sewage Pollution Manifesto, developed by a 'coalition of water lovers and water users'.[52] This shows that that the UK has a nascent water movement. What we can do in this section of the chapter is make sense of the aims and achievements of the existing movement and understand its potential for development to the point where it could become a constructively disruptive force.

In thinking about these issues, it is helpful to start from Donatella Della Porta's distinction between civil society and social movements.[53] Civil society action seeks improvements within the existing framework of policy and practice, while social movements seek to change that framework. This can be used as a classificatory device, recognising of course that any distinction of this kind is overly neat and static, when actual organisations or collaborations exist at a point on a continuum rather than at one or other pole. Moreover, any existing water movement will include fractions with different agendas seeking

Towards foundational water management

to maintain or to shift the movement's position towards one or other of the poles. After noting these complexities, it is clear that creating a social movement that seeks to change the framework of policy and practice could help redirect water management and propel it towards the foundational ideal.

On this basis, the experience of water movements in continental Europe is encouraging, because in the past decades they have been able to change water policy in various countries, upending arrangements that suited financial and political elites.[54] Social movements have emerged from the 1990s and have halted the advance of privatisation or reversed the privatisation of water and sanitation services in several European cities. These movements show that power configurations can be disrupted, the terms of the debate can be changed, and ultimately citizen-led reforms in water policy are possible. Compared with campaign groups in the UK, European water movements started with the practical advantage that European practice had been to privatise operations management but retain municipal ownership of water company assets. It is easier to reverse this kind of operating franchise privatisation than it would be to re-nationalise in the UK, where both assets and operations were privatised, and assets would have to be bought or taken back. Nonetheless, the European demands were those of a radical social movement, as we explain below.

Two key documents – the Naples Manifesto of 2012 and the 2023 Position Paper in response to the EU Blue New Deal – give a sense of the politics of this water movement. The 2012 Naples Manifesto contrasts private management of water as a commodity with democratic management of water as a commons.[55] This reframing as commons recognises a 'fundamental right' to water, which needs to take account of the 'local water cycle' and 'the rights of nature', thus connecting social justice with the need for environmental protection and regeneration. The European water movement also recognises the need for more investment, which in the 2012 manifesto is seen as a 'collective responsibility, to be paid for through general taxation'. In the 2023 Position Paper, this principle is developed further with an explicit point about progressive taxation so that water is affordable for low-income households.[56] The publicly owned company (usually municipal or regional) is reinstated but with a requirement for democratisation through citizen, worker and NGO participation.

In several major European cities, local water movements have gained enough power to negotiate with local authorities to reverse privatisation or win major concessions to their demands. Two prominent examples are Paris and Naples. In both cases innovations include the 'opening' of the water company's board of directors to community stakeholders, with the creation of a water observatory as an external advisory body with a broad democratic base. This is more or less what Labour proposed in its 2018 plan for public ownership mentioned earlier. In some places in Europe, promoted and backed by a social movement, the democratised water company has become a reality. In the UK, after Jeremy Corbyn lost the 2019 election and the centre-right recaptured the Labour leadership, the idea of democratisation was lost along with the core proposal for public ownership. This strongly suggests the limits of radical ideas without social movements that can provide sustained pressure.

The European water movement has often been close to political parties at municipal level and has used civic instruments such as the European Citizens' Initiative.[57] But the demands of this movement position it well towards the social movement end of the civil society/social movement continuum. It represents a clear break with mainstream water practice in proposing alternative forms of community management of water and justifying this in a distinctive language of rights and commons, while at the same time engaging with practical issues such as charging systems. The objective has been not only to capture control but also to democratise the water company. On the basis of these achievements, it now faces new challenges of adaptation to climate change where issues such as water cycle planning and multi-stakeholder coordination will require further innovations.

If the UK is a long way behind, over the past few years a new water movement *has been* forming outside of formal political party organisations. The activities and proposals emerging from this putative movement show great potential, though existing limitations need to be recognised and addressed. In an exemplary way, the new movement is based not just on the direct experience and voice of groups such as anglers, wild swimmers and surfers, but on producing systematic knowledge which draws aside the curtain of ignorance that results from poor official data and monitoring, and both informs and radicalises activists. For example, citizen scientists from Windrush against Sewage Pollution (WASP) have been collecting data on sewage spills since 2010, with Peter Hammond, a retired professor, providing expert input and giving cogent evidence at inquiries. This work has revealed the scale of pollution in the River Windrush specifically, and the general inadequacy of Environment Agency data, which covers only 3.5% of the total spills over that period.[58] More broadly, the investigative journalism of Watershed aims to cover 'all aspects of the water crisis: pollution, resources, over-abstraction, wildlife, public health, environmental justice, and the impacts of climate change'.[59] Their activities have included collecting data from numerous official and citizen sources on a wide range of pollution sources, as well as collaborating with *The Guardian* on investigations, such as that on 'forever chemicals'.[60]

It is not straightforward to classify the UK water movement on the civil society/social movement continuum because, like its European counterpart, it is heterogeneous and consists of groups with different and changing relations to the framework around the industry. At the civil society end of the continuum, Southend-based Waterwatch is an independent organisation trying to hold water companies and others to account through partnerships, which includes drawing funding from Anglian Water.[61] At the opposite end, Extinction Rebellion has taken up the dirty water issue and added a direct action campaign of bill boycotts under the slogan 'don't pay for dirty water'.[62] Equally importantly, long-established organisations mutate over time. The Rivers Trust was radicalised in the second half of the 2010s, with figures such as Feargal Sharkey insisting that 'our chalk streams … cannot wait'.[63] Surfers against Sewage was founded in the 1990s around a single issue but now targets many forms of ocean pollution, including plastics, and has the positive objective of 'making the ocean thrive',[64] working in a collaborative way with other organisations.

Tactics are variable and include both using the law to harass water companies and protesting against the law. There is a longstanding tradition of using existing law against the water companies which goes back at least to the Fish Legal 2012 and 2015 cases, which secured judgements that water companies were public bodies and as such were obliged to disclose information about water quality.[65] The tactic continues with River Action targeting the Environment Agency and Defra regarding runoff polluting the Wye.[66] But, equally, Surfers against Sewage, River Action and WASP are protesting about new laws in the form of the Labour government's 2024 Water Bill. They have demanded that the 'bailouts provision is removed from the bill, to stop billpayers and taxpayers having to pay off water company debts and shareholders', and ask that 'Ofwat's farcical duty to ensure water companies make a profit must be replaced with a legal duty to protect the environment and public health'.[67]

However, and this is an important qualification, when the UK water organisations are brought together on a lowest common denominator basis, their demands are very much those of a civil society movement which wants to make things work better within the existing framework by using law and regulation. This is clear from the 2023 End Sewage Pollution Manifesto, which was created by eight organisations including Surfers against Sewage, River Action, Anglers Trust and Rivers Trust.[68] The aims and demands of the manifesto were carried over directly into the one-day protest of the 2024 March for Clean Water which listed a hundred supporting organisations, including Water UK.[69]

- The overarching aim of the manifesto is cleaner water through 'making the UK's waters healthy and safe again', with no context about the larger and more challenging problem of adaptation to climate change. The aim is consensual, though nobody could be against cleaner water. And there is no sign of any radical, justificatory language about rights and commons.
- Equally consensual are the proposed solutions, where the manifesto places much emphasis on enforcing the law 'because we have the regulations and laws we need to end

sewage pollution'. With the focus on sewage pollution, broader issues such as the inequities of the existing regressive household charging system do not figure.
- The headline demand is that the water companies 'stop pollution for profit' through capping CEO bonuses and making payment of dividends conditional upon environmental performance. Again, nobody could be in favour of pollution for profit and the slogan allows the manifesto to dodge the question of public ownership of the industry, which is more overtly political.

This suggests a clear tension between the need to find a broad base of support and build collaborative networks of organisations and the need to develop and sustain a radical rethinking of water. But as we have noted, the UK water movement is heterogeneous and within that variety there is serious radicalism about problems and solutions in old and new mainstream organisations with large memberships. One good example is the Rivers Trust, an association of some sixty member trust charities concerned with one river or the rivers in a sub-region. The Rivers Trust recognises a breadth of issues including adaptation to climate change, the importance of catchment and nature-based solutions, and the problem of 'diffuse urban pollution' as well as agricultural runoff.[70] The Rivers Trust is part of the Sustainable Solutions for Water and Nature (SSWAN) partnership of organisations, which 'share the same goal to find sustainable solutions for water and nature', and advocate 'a catchment wide approach focusing on nature-based and low carbon solutions'.[71] The SSWAN partnership's discussion paper proposes a new outcomes-based regulatory framework reflecting a catchment approach to planning, in that it involves multiple actors 'with national targets set, apportioned and incentivized' partly by financial penalties for non-delivery.[72]

The SSWAN partnership is interesting because of the breadth and depth of the partner organisations it already includes. These include the Sustainability First think tank concerned with 'community and nature based solutions', which has also worked on unfair energy charging. Along with the Rivers Trust, the partnership includes another association, the Wildlife Trusts, which brings together 46 small local charities that conserve wildlife and habitats, and the RSPB, which is, like the National Trust, a subscription-based mass membership organisation with 1.2 million members. And, finally, it also includes Wessex Water, owned by the Malaysian YTL conglomerate. Wessex Water stands out as the first water company to recognise that the old game has to change in a period of nature and climate emergency. Of course, any water company will also have a vested interest in the status quo, and participation in 'partnership' arrangements can be a way of largely maintaining existing frameworks and power. It is quite possible to engage in some sustainability-related actions while not making wholesale changes to practices and priorities. Wessex Water's participation in the partnership is thus an indicator of the difficulties that broad-based partnerships are likely to face as new partners bring strength in numbers and broader representation. Partnerships should provide opportunities for experimentation and learning around solutions and collaborations, but might also compromise the radical agenda which includes replacing the private companies.

Radicalism can come from a realist recognition of the gravity of the problem of water management under climate change, as much as from an ideological preference. SSWAN has a broad problem definition, but it is not at all clear whether its members can or wish to make the transition to a politics of organised disruption which came naturally to the European movement, given its ideological basis in the language of the commons and rights and its hostility to commodification. The challenge is to see whether many more organisations can rally around more radical common denominators and put together a new kind of politics that connects planning with democracy and recognises the need to reinvent the water company for adaptation to climate change. The advantage is that the existing settlement built around private water companies is completely discredited, and public indignation about storm overflows provides an impetus which could be channelled in progressive directions. A broad base of activists in focused organisations could build the leverage to disrupt the power configuration and add external pressure for the reform of water management, not just fixing storm overflows. If that is a real possibility, the outcome is uncertain because it depends on activist education and organisation.

Cynics might respond that water is a foundational necessity, but prioritising social movements in water is a bit of a luxury in the world as it is. We live with exogenous shocks from pandemics and from geopolitical instability in a multipolar world. And then there is the internal corrosion of citizen rights that is dramatised by the authoritarian turn in the United States and the five European 'dismantler states', which now include Italy, according to the Civil Liberties Union for Europe.[73] Of course, it is important to respond sensibly to shocks and to defend liberal rights. But an effective response to outside threats and defence against internal dismantlers depends on our capacity to challenge power in our own societies by building more substantial democracies with active citizens. Creating a social movement in water which can disrupt the power configuration is an early and important part

Towards foundational water management

of these broader struggles. Citizenship is not a status nor a fixed set of rights, but an activity that is constantly renewed through struggle for participation in setting the agenda and the policy pathways towards foundational liveability and sustainability in each new conjuncture.

Notes

Introduction

1 Definition from the *Cambridge Dictionary* online, https://dictionary.cambridge.org/dictionary/english/murky#google_vignette (accessed 23 May 2025).
2 Nils Pratley, 'Thames Water on TV: pity the staff, this place is decrepit', *The Guardian*, 16 March 2025, https://www.theguardian.com/business/2025/mar/16/thames-water-on-tv-pity-the-staff-this-place-is-decrepit (accessed 23 May 2025).
3 Lena Swedlow, *Our Water Our Way: A Democratic Case for Public Ownership of Water* (2025), Compass, https://www.compassonline.org.uk/publications/our-water-our-way/ (accessed 23 May 2025).
4 Windrush, https://www.windrushwasp.org/data-analysis (accessed 23 May 2025).
5 Joe Crowley, 'Sewage dumped illegally in Windermere over 3 years', BBC News, 17 October 2024, https://www.bbc.co.uk/news/articles/cdrj70dynk1o (accessed 23 May 2025).
6 The Rivers Trust, 'The Rivers Trust response: Environmental Audit Committee's Water Quality and Water Infrastructure follow-up inquiry' (2024), evidence, p. 5, https://committees.parliament.uk/writtenevidence/130196/pdf/ (accessed 23 May 2025).
7 Environment Agency, 'Environment Agency storm overflow spill data for 2024', press release, 27 March 2024, https://www.gov.uk/government/news/environment-agency-storm-overflow-spill-data-for-2024 (accessed 23 May 2025).
8 Conservative Home, 'Kemi Badenoch: net zero by 2050 "is fantasy politics. Built on nothing. Promising the earth. And costing it too"',

Notes

18 March 2025, https://conservativehome.com/2025/03/18/kemi-badenoch-net-zero-by-2050-is-fantasy-politics-built-on-nothing-promising-the-earth-and-costing-it-too/ (accessed 23 May 2025).

9 Defra and Welsh Government, 'Governments launch largest review of sector since privatisation', press release, 22 October 2024, https://www.gov.uk/government/news/governments-launch-largest-review-of-sector-since-privatisation (accessed 23 May 2025).

10 Swedlow, *Our Water Our Way*, p. 5.

11 Foundational Economy Collective, *The Foundational Economy: The Infrastructures of Everyday Life* (Manchester University Press, 2018).

12 Luca Calafati, Julie Froud, Colin Haslam, Sukhdev Johal and Karel Williams, *When Nothing Works: From Cost of Living to Foundational Liveability* (Manchester University Press, 2023).

13 BBC News, 'How much will I pay for my water and how can I cut my bill?', 11 July 2024, https://www.bbc.co.uk/news/articles/cmm26e1qpzgo (accessed 23 May 2025).

14 Julie Froud, Sukhdev Johal, Adam Leaver and Karel Williams, *Financialisation and Strategy* (Routledge, 2006).

15 Ofwat. *Future Water and Sewerage Charges 2005–10* (2004), p. 264, https://www.ofwat.gov.uk/wp-content/uploads/2020/10/PR04-final-determinations-document.pdf (accessed 23 May 2025); Ofwat, *Future Water and Sewerage Charges 2020–15: Final Determinations* (2009), p. 128, https://www.ofwat.gov.uk/wp-content/uploads/2015/11/det_pr09_finalfull.pdf (accessed 23 May 2025); Ofwat, *Final Price Control Determination Notice: Policy Chapter A7 – Risk and Reward* (2014), p. 42, https://www.ofwat.gov.uk/wp-content/uploads/2015/10/det_pr20141212riskreward.pdf (accessed 23 May 2025).

16 Afonydd Cymru analysis indicates that in 2024 the 'top 20' storm overflows were operating legally for only 1% of the time. Afonydd Cymru, 'Welsh Water publish 2021 storm overflow data', 27 March 2025, https://afonyddcymru.org/welsh-water-publish-2024-storm-overflow-data-2/ (accessed 23 May 2025).

17 Andrew Bowman, Julie Froud, Sukhdev Johal and Karel Williams, 'Trade associations, narrative and elite power', *Theory, Culture & Society*, 34.5–6) (2017), 103–26.

18 Gill Plimmer and Clara Murray, 'Water companies fail to improve more than half of worst sewage overflow pipes', *Financial Times*, 27

Notes

March 2025, https://www.ft.com/content/de8c5104-2384-4bef-8fd5-ac58545e6903 (accessed 23 May 2025).

19 Steve Reed, 'Steve Reed speech on the Water (Special Measures) Bill', Defra, 5 September 2024, https://www.gov.uk/government/speeches/steve-reed-speech-on-the-water-special-measures-bill (accessed 23 May 2025).

20 Industry and Regulators Committee, 'Corrected oral evidence: the work of Ofwat', House of Lords, 6 September 2022, p. 14, https://committees.parliament.uk/oralevidence/10669/pdf/ (accessed 23 May 2025).

Chapter 1

1 Foundational Economy Collective, *Foundational Economy: The Infrastructure of Everyday Life* (Manchester University Press, 2018).

2 D. A. K. Black, R. A. McCance and W. F. Young, 'A study of dehydration by means of balance experiments', *The Journal of Physiology*, 102 (1944), 106–14.

3 WHO, 'Drinking water. Key facts', 13 September 2023, https://www.who.int/news-room/fact-sheets/detail/drinking-water (accessed 23 May 2025).

4 WHO, UNICEF and World Bank, *State of the World's Drinking Water* (2022), https://iris.who.int/bitstream/handle/10665/363704/9789240060807-eng.pdf?sequence=1&isAllowed=y (accessed 23 May 2025).

5 Karl August Wittfogel, *Oriental Despotism: A Comparative Study of Total Power* (Yale University Press, 1957).

6 David Kinnersley, *Troubled Water* (Hilary Shipman, 1988), pp. 75–92; Elizabeth Porter, *Water Management in England and Wales* (Cambridge University Press, 1978), pp. 27–30.

7 *The Economist*, 8 February 1986, quoted by Kinnersley, *Troubled Water*, p. 146.

8 Privatisation of the Water Authorities in England and Wales, Cmnd 9734, 1986.

9 Privatisation of the Water Authorities in England and Wales, Cmnd 9734, 1986, p. 2.

Notes

10 Faisal Islam, 'Water should be a simple business – why isn't it?', BBC News, 29 June 2023, https://www.bbc.co.uk/news/business-66056024 (accessed 23 May 2025).

11 Dieter Helm, 'A bad answer to the wrong questions: Ofwat's interim determination and its Turnaround Oversight Regime for Thames Water', blog, 15 July 2024, https://dieterhelm.co.uk/publications/a-bad-answer-to-the-wrong-questions-ofwats-interim-determination-and-its-turnaround-oversight-regime-for-thames-water/ (accessed 23 May 2025).

12 Water company business plan submissions as part of PR24, the price setting process for 2025–30.

13 Department for Transport, 'Road lengths in Great Britain: 2020', 4 February 2021, https://assets.publishing.service.gov.uk/media/601989fd8fa8f53fc149bc4b/road-lengths-in-great-britain-2020.pdf (accessed 23 May 2025).

14 Water company business plan submissions as part of PR24, the price setting process for 2025–30.

15 Sarah Tudor, 'Sewage pollution in England's waters', House of Lords Library, 30 June 2022, https://lordslibrary.parliament.uk/sewage-pollution-in-englands-waters/ (accessed 23 May 2025); Ed Conway, 'Down the drain: what went wrong with Britain's water system', Sky News, 27 October 2023, https://news.sky.com/story/down-the-drain-what-went-wrong-with-britains-water-system-12987514 (accessed 23 May 2025).

16 Ofwat, 'Leakage in the water industry', 21 November 2022, https://www.ofwat.gov.uk/leakage-in-the-water-industry/ (accessed 23 May 2025).

17 Defra, 'Governments launch largest review of sector since privatisation', press release, 22 October 2024, https://www.gov.uk/government/news/governments-launch-largest-review-of-sector-since-privatisation (accessed 23 May 2025).

18 See, for example, the National Oceanic and Atmospheric Administration: https://www.noaa.gov/education/resource-collections/freshwater/water-cycle (accessed 23 May 2025).

19 Institution of Civil Engineers, https://www.ice.org.uk/what-is-civil-engineering/infrastructure-projects/thirlmere-aqueduct-and-reservoir (accessed 23 May 2025).

Notes

20 Stephen Halliday, *The Great Stink of London* (The History Press, 2001), p. 124.
21 Grace's Guide, https://www.gracesguide.co.uk/Haweswater_Waterworks (accessed 23 May 2025).
22 Water and wastewater: https://www.waterandwastewater.com/rapid-gravity-filtration-enhancing-water-treatment-efficiency/ (accessed 23 May 2025).
23 Glen T. Daiger, 'Ardern and Locket remembrance', in D. Jenkins and J. Warner (eds), *Activated Sludge – 100 Years and Counting* (IWA Publishing, 2014), pp. 1–16.
24 Halliday, *The Great Stink*, pp. 149–53.
25 Halliday, *The Great Stink*, pp. 79, 89.
26 Institute of Civil Engineers, https://www.ice.org.uk/what-is-civil-engineering/infrastructure-projects/london-sewer-system (accessed 23 May 2025).
27 David Lewis Brown, *Elan Valley Clearance* (Fircone Books, 2019), p. 24.
28 Science & Industry Museum, 'Slums and suburbs: water and sanitation in the first industrial city', 11 February 2021, https://www.scienceandindustrymuseum.org.uk/objects-and-stories/water-and-sanitation (accessed 23 May 2025).
29 Substormflow: Manchester, https://substormflow.com/location/manchester/ (accessed 23 May 2025).
30 Halliday, *The Great Stink*, pp. 80–1.
31 Paul Dobraszczyk, *London's Sewers* (Bloomsbury, 2014), p.18.
32 Halliday, *The Great Stink*, pp. 22–3.
33 Peter Ackroyd, *Thames: Sacred River* (Vintage, 2008), p. 389.
34 Roger Milne, 'Technology: sludge disposal becomes a burning issue', *New Scientist*, 1 December 1990, p. 3, https://www.newscientist.com/article/mg12817453-700-technology-sludge-disposal-becomes-a-burning-issue/ (accessed 23 May 2025).
35 Nautilus International, https://www.nautilusint.org/en/news-insight/ships-of-the-past/2022/december/bexley/ (accessed 23 May 2025).
36 Greenpeace, https://media.greenpeace.org/archive/Sludge-Dumping-Protest-in-North-Sea-27MZIF7GX09.html (accessed 23 May 2025).
37 Brown, *Elan Valley Clearance*, p. 24.

Notes

38 Bishop of Manchester's speech, reported in *Manchester Courier and Lancashire General Advertiser*, 4 October 1879, cited https://en.wikipedia.org/wiki/Thirlmere (accessed 23 May 2025).
39 Lake District National Park, https://www.lakedistrict.gov.uk/caringfor/whs/marketing-toolkit/thirlmere (accessed 23 May 2025).
40 Brown, *Elan Valley Clearance*, p. 150.
41 Abandoned Communities (undated), 'Reservoirs of Wales', http://www.abandonedcommunities.co.uk/page32.html (accessed 23 May 2025).
42 Brown, *Elan Valley Clearance*, p. 150.
43 From the Manics 1998 album *This Is My Truth, Tell Me Yours*. Those who find the original rock version over-produced should sample the James Dean Bradfield and John Cale acoustic version which is available on YouTube. Cale, best known for his stint in the Velvet Underground, is a Welsh-speaking miner's son from Garnant. 'Blas y cyw yn y cawl' as the Welsh say.
44 Fritz Möller, 'On the influence of changes in the CO_2 concentration in air on the radiation balance of the Earth's surface and on the climate', *Journal of Geophysical Research*, 63.13 (1953), 3877–86.
45 Syukuro Manabe and Richard T. Wetherald, 'The effects of doubling the CO_2 concentration on the climate of a general circulation model', *Journal of the Atmospheric Sciences*, 32.1 (1967), 3–15; Wallace S. Broecker, 'Climatic change: are we on the brink of a pronounced global warming?', *Science*, 189.4201 (1975), 460–3.
46 United Nations Framework Convention on Climate Change (UNFCC), *The Paris Agreement* (2016), https://unfccc.int/process-and-meetings/the-paris-agreement (accessed 23 May 2025).
47 UNFCC, 'Why the Global Stocktake is important for climate action this decade', https://unfccc.int/topics/global-stocktake/about-the-global-stocktake/why-the-global-stocktake-is-important-for-climate-action-this-decade (accessed 23 May 2025).
48 Copernicus, 'The year 2024 set to end up as the warmest on record', Climate Change Service, 7 November 2024, https://climate.copernicus.eu/year-2024-set-end-warmest-record (accessed 23 May 2025).
49 Jean-Baptiste Fressoz, *More and More and More: An All-Consuming History of Energy* (Allen Lane, 2024), pp. 1–12.

Notes

50 Fressoz, *More and More and More*, p. 3.
51 Sources: Our World in Data CO_2 emissions, https://ourworldindata.org/co2-emissions (accessed 23 May 2025); World Bank GDP in constant US dollars, https://data.worldbank.org/indicator/NY.GDP.MKTP.KD?end=2023&start=1990 (accessed 23 May 2025). Note: 'CO_2e' stands for carbon dioxide equivalents.
52 Helmut Haberl, Dominik Wiedenhofer, Doris Virág, Gerald Kalt, Barbara Plank, Paul Brockway, Tomer Fishman, Daniel Hausknost, Fridolin Krausmann and Bartholomäus Leon-Gruchalski, 'A systematic review of the evidence on decoupling of GDP, resource use and GHG emissions, part II: Synthesizing the insights', *Environmental Research Letters*, 15.6 (2020), 065003.
53 Climate Change Committee, *The UK Climate Change Act*, CCC Insights Briefing 1 (2020), https://www.theccc.org.uk/wp-content/uploads/2020/10/CCC-Insights-Briefing-1-The-UK-Climate-Change-Act.pdf (accessed 23 May 2025).
54 Department for Business, Energy and Industrial Strategy, 'UK becomes first major economy to pass net zero emissions law', press release, 27 June 2019, https://www.gov.uk/government/news/uk-becomes-first-major-economy-to-pass-net-zero-emissions-law (accessed 23 May 2025).
55 Oxfam, *Tightening the Net. Net Zero Targets – Implications for Land and Food Equity* (2021), https://oxfamilibrary.openrepository.com/bitstream/handle/10546/621205/bp-net-zero-land-food-equity-030821-en.pdf;jsessionid=94FB13CCE5A7F77BD8C3589F0EBC6E93?sequence=1 (accessed 23 May 2025).
56 Reform UK, 'Our Contract With You' (2024), p. 8, https://assets.nationbuilder.com/reformuk/pages/253/attachments/original/1718625371/Reform_UK_Our_Contract_with_You.pdf?1718625371 (accessed 23 May 2025).
57 Bob Ward, 'Costs and benefits of the UK reaching net zero emissions by 2050: the evidence', LSE/Grantham Research Institute, 3 August 2023, https://www.lse.ac.uk/granthaminstitute/news/costs-and-benefits-of-the-uk-reaching-net-zero-emissions-by-2050-the-evidence/ (accessed 23 May 2025).
58 John Rodda and Terry March, 'The 1975–76 drought. A contemporary and retrospective review', Centre for Ecology & Hydrology

Notes

(2011), p. 37, https://nora.nerc.ac.uk/id/eprint/15011/1/CEH_1975-76_Drought_Report_Rodda_and_Marsh.pdf (accessed 23 May 2025).

59 Alyson Chapman, 'With climate change, global lake evaporation loss larger than previously thought', Texas A&M University News, 28 June 2022, https://engineering.tamu.edu/news/2022/06/with-changing-climate-global-lake-evaporation-loss-larger-than-previously-thought.html (accessed 23 May 2025).

60 Source: Met Office UK and regional series, www.metoffice.gov.uk/research/climate/maps-and-data/uk-and-regional-series (accessed 23 May 2025).

61 Source: Met Office UK and regional series, www.metoffice.gov.uk/research/climate/maps-and-data/uk-and-regional-series (accessed 23 May 2025).

62 Source: Met Office UK and regional series, https://www.metoffice.gov.uk/research/climate/maps-and-data/uk-and-regional-series (accessed 23 May 2025).

63 Met Office UK and regional series, https://www.metoffice.gov.uk/research/climate/maps-and-data/uk-and-regional-series (accessed 23 May 2025).

64 Peter Foster, 'Michael Gove's plan for thousands of Cambridge homes at risk from lack of water', *Financial Times*, 27 January 2024, https://www.ft.com/content/d1c0bf52-c8ed-4673-9aa3-3df6c771e7a7 (accessed 23 May 2025).

65 Tim Foster, 'No room for drought: steps to improve the UK agricultural sector's resilience to drought and water security', *Policy@Manchester*, blog, 10 January 2024, https://blog.policy.manchester.ac.uk/energy_environment/2024/01/no-room-for-drought-steps-to-improve-the-uk-agricultural-sectors-resilience-to-drought-and-water-security/ (accessed 23 May 2025).

66 Met Office (undated), 'UK and global extreme events – heavy rainfall and floods', https://www.metoffice.gov.uk/research/climate/understanding-climate/uk-and-global-extreme-events-heavy-rainfall-and-floods (accessed 23 May 2025).

67 Julia Bryson and Seb Cheer, 'Flood threat has "devastating" impact, say traders', BBC News, 26 November 2024, https://www.bbc.co.uk/news/articles/cew2e29kwqjo (accessed 23 May 2025).

Notes

68 Antonia Matthews and Paul Pigott, 'Anger over level and timing of storm weather warnings', BBC News, 25 November 2024, https://www.bbc.co.uk/news/articles/cgj7qvj097po (accessed 23 May 2025).

69 Environment Agency, *National Assessment of Flood and Coastal Erosion Risk in England 2024* (2024), https://assets.publishing.service.gov.uk/media/6797a4e6e0edc3fbb060633c/E03253099_EA_Flood_Coastal_Erosion_Risk_Assessment_accessible_v2.pdf (accessed 23 May 2025).

Chapter 2

1 Gill Plimmer and Robert Smith, 'Thames Water faces fresh legal challenge over £3bn creditor loan', *Financial Times*, 21 January 2025, https://www.ft.com/content/25b857d2-0650-4f48-8cfa-f63c1096918f (accessed 23 May 2025).

2 Hayley Dunning, 'Sewage overspills result from lack of infrastructure investment, research shows', Imperial News, 30 January 2023, https://www.imperial.ac.uk/news/242831/sewage-overspills-result-from-lack-infrastructure/#comments (accessed 23 May 2025).

3 Gill Plimmer and Clara Murray, 'Most of Thames Water's sewage plants unable to cope with demand', *Financial Times*, 13 February 2025, https://www.ft.com/content/21ec8e46-5864-42cf-9840-7fdd36f810ce (accessed 23 May 2025).

4 David Kinnersley, *Troubled Water: Rivers, Politics and Pollution* (Hilary Shipman, 1988), p. 135.

5 Matthew Smith, 'Support for nationalising utilities and public transport has grown significantly in last seven years', YouGov, 18 July 2024, https://yougov.co.uk/politics/articles/50098-support-for-nationalising-utilities-and-public-transport-has-grown-significantly-in-last-seven-years (accessed 23 May 2025).

6 Jean Shaoul, 'A critical financial analysis of the performance of privatised industries: the case of the water industry in England and Wales', *Critical Perspectives on Accounting*, 8 (1997), 479–505.

7 Emanuele Lobina and David Hall, 'UK water privatisation – a briefing', Public Services International Research Unit,

Notes

University of Greenwich (2001), https://drive.google.com/file/d/1655Co09hAd4yiEQ6wWni0753B7EpA1Jx/view?pli=1 (accessed 23 May 2025).

8 David Hall and Emanuele Lobina, *Clean Water: A Case for Public Ownership*, PSIRU report for Unison (2024), https://gala.gre.ac.uk/id/eprint/47676/7/47676%20LOBINA_Clean_Water_A_Case_For_Public_Ownership_2024.pdf (accessed 23 May 2025).

9 Kate Bayliss, 'Case study: the financialisation of water in England and Wales', FESSUD Working Paper no. 52 (2014), https://soas-repository.worktribe.com/output/339596 (accessed 23 May 2025).

10 John Allen and Michael Pryke, 'Financialising household water: Thames Water, MEIF, and the "ring-fenced" politics', *Cambridge Journal of Regions, Economy and Society*, 6 (2013), 419.

11 Dieter Helm, 'Whither water regulation', in D. R. Helm (ed.), *Water, Sustainability and Regulation* (The Oxera Press, 2003); Dieter Helm and Tom Tindall, 'The evolution of infrastructure and utility ownership and its implications', *Oxford Review of Economic Policy*, 25.3 (2009), 425–6.

12 Kate Bayliss and David Hall, 'Bringing water into public ownership: costs and benefits' (2017), https://gala.gre.ac.uk/id/eprint/17277/3/17277%20HALL_Bringing_Water_into_Public_Ownership_2017.pdf (accessed 23 May 2025).

13 Bayliss and Hall, 'Bringing water into public ownership', p. 3.

14 Bayliss and Hall, 'Bringing water into public ownership', p. 4.

15 The data presented in this chapter covers the ten integrated water supply and sewerage treatment companies in England and Wales, which collectively represent around 97% of the industry's revenue. As of December 2024 there are also 14 smaller companies that provide only water supply services, and these entities are not considered in the analysis presented in this book. Additionally, Scotland and Northern Ireland are omitted from the analysis as they function outside the regulatory framework established by Ofwat.

16 Source: Company annual report and accounts 1989–2023, and UK CPI, ONS. Note: data is adjusted for inflation and presented in 2023 prices.

17 Michael Gove, 'A water industry that works for everyone', 1 March 2018, https://www.gov.uk/government/speeches/a-water-industry-that-works-for-everyone (accessed 23 May 2025).

Notes

18 Gill Plimmer and Ella Hollowood, 'Sewage spills highlight decades of under-investment at England's water companies', *Financial Times*, 28 December 2021, https://www.ft.com/content/b2314ae0-9e17-425d-8e3f-066270388331 (accessed 23 May 2025).

19 Gove, 'A water industry that works for everyone'.

20 Ofwat sets water charges and investment levels on a five-year price review (PR) cycle. PR19 covered the period 2020–25.

21 Gove, 'A water industry that works for everyone'.

22 Source: Company annual report and accounts, various years, and Fame, BvDep. Notes: data is the average between 2012 to 2023. Physical assets are also known as tangible assets.

23 This is similar to private finance initiative (PFI) projects used to finance and build hospitals, schools, prisons, roads and a range of other public infrastructures in the 1990s and 2000s.

24 Chris Ames, '38% of Britain's rail network now electrified', *The Transport Network*, 5 November 2020, https://www.transport-network.co.uk/38-of-Britains-rail-network-now-electrified/16944 (accessed 23 May 2025).

25 Katy Austin, 'Motorway electric car charge point target missed, says RAC', BBC News, 2 January 2024, https://www.bbc.co.uk/news/business-67858961 (accessed 23 May 2025).

26 Andrew Bowman, Ismail Erturk et al., *The End of the Experiment* (Manchester University Press, 2014), p. 35.

27 David S. Saal and David Parker, 'Productivity and price performance in the privatized water and sewerage companies of England and Wales', *Journal of Regulatory Economics*, 20.1 (2001), 61–90.

28 Frontier Economics, *Productivity Improvement in the Water and Sewerage Industry in England Since Privatisation*, Final Report for Water UK (2017), https://www.water.org.uk/sites/default/files/wp/2018/11/Water-UK-Frontier-Productivity.pdf (accessed 23 May 2025).

29 Source: BT annual report and accounts, 1985 (the first year after privatisation).

30 Stuart G. Ogden, 'Transforming frameworks of accountability: the case of water privatization', *Accounting, Organizations and Society*, 23.2/3 (1995), 196.

31 Shaoul, 'A critical financial analysis of the performance of privatised industries', pp. 491–2.

Notes

32 Annual report and accounts of the England and Wales water companies.
33 Operating cashflow is the cash received from water companies' customers minus the costs of labour and other operating expenses. It is approximately equivalent to EBITDA: earnings (profit) before interest, tax, depreciation and amortisation.
34 Julian Hofmann, 'National Grid rights issue secures the balance sheet', *Investors Chronicle*, 7 November 2024, https://www.investorschronicle.co.uk/content/6250d654-dbbf-50fd-8b2c-4def4b3b3167 (accessed 23 May 2025).
35 Source: Company annual report and accounts, various years, and Fame, BvDep. Notes: the data is the average between 2012 and 2023. Network Rail data includes government grants. All the companies were formerly state-owned, with the water companies, National Grid and BT privatised and Network Rail created in 2002 as the successor to Railtrack.
36 Karen Bakker, *An Uncooperative Commodity: Privatizing Water in England and Wales* (Oxford University Press 2003), p. 8.
37 Privatisation of the Water Authorities in England and Wales, Cmnd 9734, 1986, p. 1.
38 Ofwat, 'Final determinations' (2019), https://www.ofwat.gov.uk/regulated-companies/price-review/2019-price-review/final-determinations/ (accessed 23 May 2025).
39 Water UK, 'Thirty years on, what has water privatisation achieved?', 18 July 2019, https://www.water.org.uk/news-views-publications/views/thirty-years-what-has-water-privatisation-achieved (accessed 23 May 2025).
40 Ofwat, 'What the 2024 price review means for water customers' (2024), https://www.ofwat.gov.uk/regulated-companies/price-review/2024-price-review/what-it-means-for-customers-and-water-bills/#pricereview2024 (accessed 23 May 2025).
41 Steve Reed, 'Price Review 2024 Final Determinations', written statement to UK Parliament, 19 December 2024, https://questions-statements.parliament.uk/written-statements/detail/2024-12-19/hcws345/ (accessed 23 May 2025).
42 David Black, Ofwat CEO, interviewed on BBC *Today* programme, 8 October 2024.

Notes

43 David Henderson, Water UK CEO, interviewed on BBC *Today* programme, 8 October 2024.

44 House of Lords Industry and Regulators Committee, *The Affluent and the Effluent: Cleaning up the Failures of Water Regulation*, 2022–23 HL Paper 166 (2023), ch. 3, paras 53–60, https://publications.parliament.uk/pa/ld5803/ldselect/ldindreg/166/16602.htm (accessed 23 May 2025).

45 House of Lords Industry and Regulators Committee, *The Affluent and the Effluent: Cleaning up the Failures of Water Regulation*, 2022–23, para 52.

46 House of Lords Industry and Regulators Committee, *The Affluent and the Effluent: Cleaning up the Failures of Water Regulation*, 2022–23, para 61.

47 House of Lords Industry and Regulators Committee, *The Affluent and the Effluent: Cleaning up the Failures of Water Regulation*, 2022–23, para 63.

48 Data extracted from PR24 company business plans, main data tables, table RR15, giving the share of revenue for PR24 (2025–30) allocated to residential households.

49 The number of dwellings in England and Wales is not the same as the number of households: a dwelling can contain more than one household, for example in properties where some facilities may be shared between more than one household. However, the discrepancy between the two is not large. For example, in 2011 there were 24.4 million dwellings and 23.4 million households in England and Wales; in 2021 there were 26.4 million dwellings and 24.8 million households, with the difference between the two totals less than 7%. Source: Table 104: by tenure, England (historical series); Table 106 Dwelling stock: by tenure, Wales (historical series) – 1990–2017 Department for Levelling Up, Housing and Communities, and Dwelling stock estimates by year and tenure, StatsWales. https://assets.publishing.service.gov.uk/media/649c2d3f06179b000c3f7426/LT_104.ods; https://assets.publishing.service.gov.uk/media/5ce6a1deed915d247979f95b/LT_106.xls (accessed 23 May 2025). Note: 2023 estimated from Housing supply: net additional dwellings, England: 2022 to 2023, ONS and 2023 for Wales data is from 'New dwellings completed by period and tenure', StatsWales.

Notes

50 Sources: Dwelling data for England from Table 104: by tenure, England (historical series), Department for Levelling Up, Housing and Communities; Wales dwelling data from 'Dwelling stock estimates by year and tenure', StatWales. Company annual report and accounts, various years. Consumer Price Index used for inflation adjustment.

51 The CPI (Consumer Price Index) is an official measure of price inflation used by the Office of National Statistics. It does not include housing costs such as mortgage payments and council tax which are included in the RPI (Retail Price Index).

52 Sources: Company annual report and accounts, various years. Note: the shading in bars represent PR periods when Ofwat set five-year targets for regulated water companies in England and Wales. Consumer Price Index used for inflation adjustment and the data is in 2023 prices.

53 See, for example, D. A. Lloyd Owen, 'Water privatization in England & Wales: past reasons, present problems and future prospects', *International Journal of Water Resources Development*, 40.6 (2024), 1078–93. On the EU Water Framework Directive, see David Woods and Neil Tytler, *The European Community Water Framework Directive* (2010), The Institute of Environmental Sciences, https://fwr.org/publication/the-european-community-water-framework-directive-an-introductory-guide/ (accessed 23 May 2025).

54 Water UK, 'Six water companies appeal to Competition and Markets Authority', press release, 18 February 2025, https://www.water.org.uk/news-views-publications/news/six-water-companies-appeal-competition-and-markets-authority (accessed 23 May 2025).

55 Privatisation of the Water Authorities in England and Wales, Cmnd 9734, 1986, p. 1.

56 *Water Briefing*, 'South West Water customers opt for shares under new ownership scheme', 3 November 2020, https://waterbriefing.org/home/company-news/item/17753-south-west-water-customers-opt-for-shares-under-new-ownership-scheme (accessed 23 May 2025). Customers were offered the choice of a £20 credit on their next bill or the equivalent in shares in Pennon Group, the owner of South West Water.

57 Source: company annual report and accounts, various years.

Notes

58 George Eustice, interview, BBC *Today* programme, 12 September 2024.
59 Nils Pratley, 'The best long term plan for Thames Water is to get it back on the stock market', *The Guardian*, 11 July 2024, https://www.theguardian.com/business/nils-pratley-on-finance/article/2024/jul/11/the-best-long-term-plan-for-thames-water-is-to-get-it-back-on-the-stock-market (accessed 23 May 2025); see also Nils Pratley, 'Here's a bright idea to improve accountability in the water sector', *The Guardian*, 3 October 2022, https://www.theguardian.com/business/nils-pratley-on-finance/2022/oct/03/bright-idea-improve-accountability-britain-water-sector (accessed 23 May 2025).
60 House of Lords Industry and Regulators Committee, Oral Evidence, 'The work of Ofwat', 6 September 2022, Q.79, https://committees.parliament.uk/oralevidence/10669/html/ (accessed 23 May 2025).
61 The Ferret, 'Claim that Scottish Water is in public hands is mostly true', 1 December 2017, https://theferret.scot/scottish-water-public-ownership/ (accessed 23 May 2025).
62 WICS, *Scottish Water's Performance 2022–23* (2024), p. 14, https://wics.scot/publications/scottish-water/performance/scottish-water-performance-report-2022-23 (accessed 23 May 2025).
63 Labour Party, *Clear Water* (2018), https://www.labour.org.uk/wp-content/uploads/2018/09/Conference-2018-Water-pamphlet-FINAL.pdf (accessed 23 May 2025).
64 Bayliss and Hall, 'Bringing water into public ownership', p. 5.
65 Gill Plimmer, 'Water renationalisation to cost as little as £14.5bn', *Financial Times*, 26 April 2019, https://www.ft.com/content/8ee5d48a-6103-11e9-a27a-fdd51850994c (accessed 23 May 2025).
66 BBC News, 'Why doesn't Northern Ireland have water bills?', 27 March 2007, http://news.bbc.co.uk/1/hi/magazine/6498983.stm (accessed 23 May 2025). A consultation exercise was held in 2023–24 on 'options for revenue raising' via water and sewerage charges, after which 'the Minister has publicly stated that he will not be introducing domestic water charges'. Department for Infrastructure, 5 July 2024, https://www.infrastructure-ni.gov.uk/publications/consultation-report-water-and-sewerage-charges-options-revenue-raising (accessed 23 May 2025).

Notes

67 Source: Company annual report and accounts, various years. The data for each company or group of companies covers the time period over which they have been in the relevant form of ownership.

68 Allen and Pryke, 'Financialising household water'; Kate Bayliss, Elisa Van Waeyenberge and Benjamin O. L. Bowles, 'Private equity and the regulation of financialised infrastructure: the case of Macquarie in Britain's water and energy networks', *New Political Economy*, 28.2 (2022), 155–72.

69 Macquarie, *Thames Water – Factsheet* (2023), p. 3, https://www.macquarie.com/assets/macq/impact/case-studies/thames-water-upgrading-londons-water-and-wastewater-infrastructure/macquarie-thames-water-factsheet.pdf (accessed 23 May 2025).

70 Source: Company annual report and accounts, various years. RWE and Macquarie ownership of Thames Water is indicated on the graph. In 2016 Thames Water implemented international accounting standards (IFRS), which resulted in a rise in the value of its infrastructure through the fair value method. This adjustment correspondingly boosted shareholder equity. Therefore, the increase in equity for 2016, as illustrated in Exhibit 2.8, is attributed to an accounting modification rather than new equity contributions from investors. For further details, see the Thames Water Utilities Limited filing history on Companies House, specifically the 2016 accounts, note 22 on page 131. For more information on the Thames Water corporate structure, see Exhibit 5.1 in Chapter 5.

71 Ofwat, 'Monitoring Financial Resilience report 2022–23 charts and underlying data', 26 October 2023, https://www.ofwat.gov.uk/publication/monitoring-financial-resilience-report-2022-23-charts-and-underlying-data/ (accessed 23 May 2025).

72 Dŵr Cymru, annual report and accounts.

73 Kuishuang Feng et al., 'Spatially explicit analysis of water footprints in the UK', *Water*, 3.1 (2011), 55, https://www.mdpi.com/2073-4441/3/1/47 (accessed 23 May 2025).

74 Waterplus (no date), 'Understanding wholesaler price changes in England for 20225/26', https://www.water-plus.co.uk/news/20252026-wholesaler-price-changes/ (accessed 23 May 2025).

75 David Mytton, 'Data centre water consumption', *npj Clean Water*, 4.11 (2021).

Notes

76 Privatisation of the Water Authorities in England and Wales, Cmnd. 9734, 1986, p. 1.
77 Defra, *Water Bill. Reform of the Water Industry: Retail Competition* (2013), briefing note, p. 1.
78 Georgina Hutton, *Water: Non-household Retail Competition*, House of Commons Library Briefing Paper CBP 8925, 29 May 2020, p. 2, https://researchbriefings.files.parliament.uk/documents/CBP-8925/CBP-8925.pdf (accessed 23 May 2025).
79 Hutton, *Water: Non-household Retail Competition*, p. 2.
80 Martin Cave, *Independent Review of Competition and Innovation in Water Markets: Final Report* (2009), Defra, https://assets.publishing.service.gov.uk/media/5a79a4ba40f0b63d72fc7633/cave-review-final-report.pdf (accessed 23 May 2025).
81 Sara Priestly and David Hough, *Increasing Competition in the Water Industry*, House of Commons Library Briefing Paper CBP 7259, 21 November 2016, pp. 8–9.
82 Ofwat, *Open for Business: Reviewing the First Year of the Business Retail Water Market* (2018), https://www.ofwat.gov.uk/wp-content/uploads/2018/07/State-of-the-market-report-2017-18-FINAL.pdf (accessed 23 May 2025).
83 Ofwat, *Business Retail Market Update 2023–24* (2024), https://www.ofwat.gov.uk/publication/business-retail-market-update-2023-24/ (accessed 23 May 2025).
84 NAO, *Water Supply and Demand Management*, 2019-21 HC 107 (2020), https://www.nao.org.uk/wp-content/uploads/2020/03/Water-supply-and-demand-management.pdf (accessed 23 May 2025).
85 Ofwat, *Costs and Benefits of Introducing Competition to Residential Customers in England* (2016), https://www.ofwat.gov.uk/wp-content/uploads/2016/09/pap_tec20160919RRRfinal.pdf (accessed 23 May 2025).
86 Hutton, *Water: Non-household Retail Competition*, p. 5.
87 Cunliffe Commission, *Call for Evidence*, Independent Commission on the Water Sector Regulatory System, 27 February 2025, pp. 150–3, https://consult.defra.gov.uk/independent-water-commission/independent-commission-on-the-water-sector-regulat/supporting_documents/Call%20For%20Evidence%20%20Independent%20Commission%20on%20the%20Water%20Sector%20Regulatory%20System.pdf (accessed 23 May 2025).

Notes

Chapter 3

1. Thames Water (no date), 'Who we are', https://www.thameswater.co.uk/about-us (accessed 23 May 2025).
2. Gill Plimmer, 'UK water companies embrace PFI to deliver £14bn of infrastructure', *Financial Times*, 17 May 2024, https://www.ft.com/content/980821ed-d6a6-4898-8ac9-4cba4ec67623 (accessed 23 May 2025).
3. Esme Stallard and Jonah Fisher, 'England sewage spills hit record 3.6m hours last year', BBC News, 27 March 2025, https://www.bbc.co.uk/news/articles/c201rz925nyo (accessed 23 May 2025).
4. Thames Water, *TMS15 Asset Health Deficit* (2023), p. 6, https://www.thameswater.co.uk/media-library/home/about-us/regulation/our-five-year-plan/pr24-2023/asset-deficit.pdf (accessed 23 May 2025).
5. Anna Isaac, 'Poor state of Thames Water a "critical risk" to UK, Starmer and Reeves told', *The Guardian*, 8 July 2024, https://www.theguardian.com/business/article/2024/jul/08/starmer-reeves-briefed-critical-risk-thames-water-whitehall-debt-infrastructure (accessed 23 May 2025).
6. Anna Isaac, 'Floods, explosions and asbestos: Thames Water faces potential problems on all fronts', *The Guardian*, 17 November 2024, https://www.theguardian.com/business/2024/nov/17/floods-explosions-asbestos-thames-water-faces-potential-problems-on-all-fronts (accessed 23 May 2025).
7. Anna Isaac, 'Thames Water's IT "falling apart" and is hit by cyber-attacks, sources claim', *The Guardian*, 18 November 2024, https://www.theguardian.com/business/2024/nov/18/thames-waters-it-falling-apart-and-is-hit-by-cyber-attacks-sources-claim (accessed 23 May 2025).
8. Sandra Laville, 'Thames Water criticised over lack of investment in sewage treatment works', *The Guardian*, 11 January 2023, https://www.theguardian.com/environment/2023/jan/11/thames-water-criticised-lack-investment-sewage-treatment-works (accessed 23 May 2025).
9. Laville, 'Thames Water criticised over lack of investment in sewage treatment works'.

Notes

10 Ofwat, 'Long-term data series of company costs' (2022), https://www.ofwat.gov.uk/publication/long-term-data-series-of-company-costs/ (accessed 23 May 2025).

11 Ofwat, 'Opex capex split' datafile (sheet F-Inputs) (2012), https://view.officeapps.live.com/op/view.aspx?src=https%3A%2F%2Fwww.ofwat.gov.uk%2Fwp-content%2Fuploads%2F2019%2F12%2FOpex-capex-split_FD.xlsx&wdOrigin=BROWSELINK (accessed 23 May 2025); Ofwat, 'Investment in the water industry', 11 March 2022, https://www.ofwat.gov.uk/investment-in-the-water-industry/#qu1 (accessed 23 May 2025).

12 Source: Long-term time series of company costs, Ofwat (online spreadsheet), https://view.officeapps.live.com/op/view.aspx?src=https%3A%2F%2Fwww.ofwat.gov.uk%2Fwp-content%2Fuploads%2F2022%2F11%2FLong-term-data-series-v4-July-2024-publication.xlsx&wdOrigin=BROWSELINK (accessed 23 May 2025). In 2015, after the introduction of total expenditure (totex), numerous companies started to classify specific costs, such as those related to infrastructure renewal, as operating expenditure (opex) rather than capital expenditure (capex), which had been the standard approach before. To facilitate consistent comparisons both among various companies and over time, this renewal expenditure has been reclassified and now falls under capex. The data also includes smaller, mainly independent water companies.

13 Ofwat, *Water Company Performance Report 2023–24* (2024), p. 5, https://www.ofwat.gov.uk/wp-content/uploads/2024/10/WCPR-23-24.pdf (accessed 23 May 2025).

14 Water mains and sewers renewed/refurbished data is sourced from business plans for the ten major water and wastewater companies supplied to Ofwat for PR24. The relevant worksheets employed to calculate water mains and sewers renewal and refurbishments are CW6 for water mains and CWW6 for the sewer network.

15 Ofwat (no date), *Business Plans*, https://www.ofwat.gov.uk/regulated-companies/price-review/2024-price-review/business-plans/ (accessed 23 May 2025). With regard to kilometres of water mains and sewers new and refurbished during the PR24 period we use the company plans covering 2025/26 and 2029/30.

Notes

16 Ofwat, *PR24 Final Determinations: Expenditure Allowances* (2024), p. 37, https://www.ofwat.gov.uk/wp-content/uploads/2024/12/PR24-final-determinations-Expenditure-allowances-V2.pdf (accessed 23 May 2025).

17 Source: Company PR24 business plans. Notes: our calculations were derived from worksheets CW1 and CWW1. The water and sewer mains mainly comprise pipes that are crucial to the water and sewage systems.

18 Ofwat, *PR24 Final Determinations: Expenditure Allowances*, p. 33.

19 T. Giakoumis and N. Voulvoulis, 'Combined sewer overflows: relating event duration monitoring data to wastewater systems' capacity in England', *Environmental Science: Water Research & Technology*, 9 (2023), 707–22.

20 Ofwat, *PR24 Draft Determinations: Expenditure Allowances – Enhancement Cost Modelling Appendix* (2024), https://www.ofwat.gov.uk/wp-content/uploads/2024/07/PR24-draft-determinations-Expenditure-allowances-Enhancement-cost-modelling-appendix.pdf (accessed 23 May 2025).

21 Ofwat, *PR24 Final Determinations: Expenditure Allowances*, pp. 15–16.

22 Ofwat, *Water Company Performance Report 2023–24*, p. 33.

23 Ofwat, *Water Company Performance Report 2023–24*, pp. 12–17.

24 Ofwat, *Water Company Performance Report 2023–24*, p. 7.

25 Ofwat, *PR24 Draft Determinations: Expenditure Allowances* (2024), p. 34, https://www.ofwat.gov.uk/wp-content/uploads/2024/07/PR24-draft-determinations-Expenditure-allowances-to-upload.pdf (accessed 23 May 2025).

26 CH2M, *Targeted Review of Asset Health and Resilience in the Water Industry* (2017), p. 9, https://www.ofwat.gov.uk/wp-content/uploads/2017/09/Targeted-Review-of-Asset-Health.pdf (accessed 23 May 2025).

27 CH2M, *Targeted Review of Asset Health and Resilience in the Water Industry* (2017), p. 8.

28 CH2M, *Targeted Review of Asset Health and Resilience in the Water Industry* (2017), pp. 2–3.

29 Andrew Bowman, Julie Froud, Sukhdev Johal and Karel Williams, 'Trade associations, narrative and elite power', *Theory, Culture & Society*, 34.5–6 (2017), 103–26.

Notes

30 Water UK, *National Storm Overflows Plan for England* (2024), https://www.water.org.uk/sites/default/files/2024-03/WEB_Water%20UK%20National%20Storm%20Overflows%20Plan%20for%20England_0.pdf (accessed 23 May 2025).

31 Water UK, *National Storm Overflows Plan for England*, p. 14.

32 Water UK, *National Storm Overflows Plan for England*, p. 15. Note: data is adapted by the authors from the chart on p. 15.

33 Water UK, *National Storm Overflows Plan for England*, p. 16.

34 Ofwat, *PR24 Draft Determinations: Expenditure Allowances*, p. 5.

35 Water UK, 'Water and sewage companies in England apologise for sewage spills and launch massive transformation programme', 18 May 2023, https://www.water.org.uk/news-views-publications/news/water-and-sewage-companies-england-apologise-sewage-spills-and-launch (accessed 23 May 2025).

36 Environment Agency, 'Storm overflow spill data shows performance is totally unacceptable', 31 March 2023, https://environmentagency.blog.gov.uk/2023/03/31/storm-overflow-spill-data-shows-performance-is-totally-unacceptable/ (accessed 23 May 2025).

37 Environment Agency, 'Storm overflow spill data shows performance is totally unacceptable', p. 5.

38 Data for 2025–30 is sourced from the PR24 business plan submissions, Worksheet CW5. Data for 2023–24 is from Water Resource Management Plan Annual Review, https://www.data.gov.uk/dataset/87b59684-3da3-45cf-8881-e4727cfd1415/water-resource-management-plan-annual-review-data (accessed 23 May 2025).

39 Water UK, *National Storm Overflows Plan for England*, p. 5.

40 National Engineering Policy Centre, *Testing the Waters: Priorities for Mitigating Health Risks from Wastewater Pollution* (2024), p. 4, https://nepc.raeng.org.uk/media/qi2eyivp/testing-the-waters-priorities-for-mitigating-health-risks-from-wastewater-pollution.pdf (accessed 23 May 2025).

41 Environment Agency, 'State of the water environment indicator B3: supporting evidence', 17 May 2024, https://www.gov.uk/government/publications/state-of-the-water-environment-indicator-b3-supporting-evidence/state-of-the-water-environment-indicator-b3-supporting-evidence (accessed 23 May 2025).

Notes

42 Jason Arunn Murugesu, 'The UK's official swimming rivers are too polluted to swim in', *New Scientist*, 8 March 2023, https://www.newscientist.com/article/2363389-the-uks-official-swimming-rivers-are-too-polluted-to-swim-in/ (accessed 23 May 2025).

43 House of Commons Environmental Audit Committee, *Water Quality in Rivers: Fourth Report of Session 2021–22* (2022), pp. 31–47, https://committees.parliament.uk/publications/8460/documents/88412/default/ (accessed 23 May 2025).

44 House of Commons Environmental Audit Committee, *Water Quality in Rivers, 2021–22 HC 74* (2022), pp. 34–5.

45 Christian M. (undated), 'Fertiliser pollution in rivers', https://www.aquaswitch.co.uk/blog/fertiliser-pollution-in-rivers/ (accessed 23 May 2025).

46 River Action (undated), 'A plan to save the Wye', https://riveractionuk.com/campaign/rescue-the-river-wye/ (accessed 23 May 2025).

47 Jo Bradley, Alasdair Chisholm and Catherine Moncrieff, *Highway Runoff and the Water Environment*, Stormwater Shepherds/CIWEM (2024), p. 13, https://mcusercontent.com/69dac29721f10aed6bd8f0af2/files/013f8803-17ac-a10e-1407-4feeb54846da/Highway_runoff_and_the_water_environment_report_combined_LR.pdf (accessed 23 May 2025).

48 Bradley, Chisholm and Moncrieff, *Highway Runoff and the Water Environment*, p. 65.

49 Bradley, Chisholm and Moncrieff, *Highway Runoff and the Water Environment*, p. 33.

50 Bradley, Chisholm and Moncrieff, *Highway Runoff and the Water Environment*, p. 78.

51 Environment Agency, *Abandoned Mines and the Water Environment* (2008), p. v, https://assets.publishing.service.gov.uk/media/5a7c3330e5274a25a9141208/LIT_8879_df7d5c.pdf (accessed 23 May 2025).

52 Laura Hughes, 'The unseen dangers of lead contamination in the UK', *Financial Times*, 6 June 2024, https://www.ft.com/content/5bcc846c-9858-4ae3-ae75-06fc17342d3d (accessed 23 May 2025).

53 Ofwat, 'Opex capex split' datasheet, 28 November 2012, https://view.officeapps.live.com/op/view.aspx?src=https%3A%2F%2Fwww.

Notes

ofwat.gov.uk%2Fwp-content%2Fuploads%2F2019%2F12%2FO pex-capex-split_FD.xlsx&wdOrigin=BROWSELINK (accessed 23 May 2025).

54 Ofwat, *Our Draft Determinations for the 2024 Price Review* (2024), Sector Summary, https://www.ofwat.gov.uk/wp-content/uploads/2024/07/PR24-DD-sector-summary.pdf (accessed 23 May 2025).

55 Water UK, *National Storm Overflows Plan for England*, p. 17.

56 The data presented in this section measures precipitation, which includes snow, sleet and hail as well as rain.

57 Deb Chachra, *How Infrastructure Works* (Penguin, 2023).

58 We use the Met Office projections based on the Representative Concentration Pathway (RCP) 8.5 (median probability), which is the higher 'business as usual' projection (including no policy mitigation) for future concentrations of greenhouse gases. See also https://www.carbonbrief.org/explainer-the-high-emissions-rcp8-5-global-warming-scenario/ (accessed 23 May 2025).

59 IPCC, 'Summary for policymakers', in *Climate Change 2021: The Physical Science Base. Contribution of Working Group 1 to the Sixth Assessment of the Intergovernmental Panel on Climate Change* (2021), https://www.ipcc.ch/report/ar6/wg1/chapter/summary-for-policymakers/ (accessed 23 May 2025).

60 Source: Met Office UK Climate Projections (2022 version), https://view.officeapps.live.com/op/view.aspx?src=https%3A%2F%2Fwww.metoffice.gov.uk%2Fbinaries%2Fcontent%2Fassets%2Fmetofficegovuk%2Fpdf%2Fresearch%2Fukcp%2Fukcp18-key-results.xlsx&wdOrigin=BROWSELINK (accessed 23 May 2025). Note: the forecasts shown in Exhibits 3.4a, 3.4b and 3.5 are based on the Representative Concentration Pathway (RCP) 8.5 (median probability). This pathway reflects a higher 'business as usual' scenario, which assumes that there are no policy measures in place to reduce greenhouse gas emissions. The average for the baseline period is calculated from 1981 to 2000.

61 Source: Met Office UK Climate Projections (2022 version), https://view.officeapps.live.com/op/view.aspx?src=https%3A%2F%2Fwww.metoffice.gov.uk%2Fbinaries%2Fcontent%2Fassets%2Fmetofficegovuk%2Fpdf%2Fresearch%2Fukcp%2Fukcp18-key-results.xlsx&wdOrigin=BROWSELINK (accessed 23 May 2025).

Notes

62 National Infrastructure Commission, *Preparing for a Drier Future: England's Water Infrastructure Needs* (2018), https://webarchive.nationalarchives.gov.uk/ukgwa/20250311123952mp_/https:/nic.org.uk/app/uploads/NIC-Preparing-for-a-Drier-Future-26-April-2018.pdf (accessed 23 May 2025); *Reducing the Risk of Surface Water Flooding* (2022), https://webarchive.nationalarchives.gov.uk/ukgwa/20250327100141/https://nic.org.uk/studies-reports/reducing-the-risks-of-surface-water-flooding/surface-water-flooding-final-report/ (accessed 23 May 2025).

63 Environment Agency, *Meeting our Future Water Needs: a National Framework for Water Resources* (2020), https://assets.publishing.service.gov.uk/media/5e6e478ed3bf7f26963789f3/National_Framework_for_water_resources_main_report.pdf (accessed 23 May 2025).

64 Environment Agency, *Meeting our Future Water Needs*, p. 7.

65 National Infrastructure Commission, *Preparing for a Drier Future*.

66 Source: Met Office UK Climate Projections (2022 version), https://view.officeapps.live.com/op/view.aspx?src=https%3A%2F%2Fwww.metoffice.gov.uk%2Fbinaries%2Fcontent%2Fassets%2Fmetofficegovuk%2Fpdf%2Fresearch%2Fukcp%2Fukcp18-key-results.xlsx&wdOrigin=BROWSELINK (accessed 23 May 2025).

67 Environment Agency, *Meeting our Future Water Needs*.

68 Environment Agency, *A Summary of England's Revised Draft Regional and Water Resources Management Plans* (2024), p. 4, https://www.gov.uk/government/publications/a-review-of-englands-draft-regional-and-water-resources-management-plans/a-summary-of-englands-draft-regional-and-water-resources-management-plans (accessed 23 May 2025).

69 Environment Agency, *Meeting our Future Water Needs*, p. 7.

70 Environment Agency, *A Summary of England's Revised Draft Regional and Water Resources Management Plans*, section 3.1.

71 Environment Agency, *Meeting our Future Water Needs*, p. 77.

72 Ofwat, 'Leakage in the water industry', 21 November 2022, https://www.ofwat.gov.uk/leakage-in-the-water-industry/ (accessed 23 May 2025).

73 Environment Agency, *Meeting our Future Water Needs*, p. 77.

74 For a summary of the regional plans produced in 2023, see https://wre.org.uk/wp-content/uploads/2023/01/SummaryofRegional

Notes

PlansforWaterResources-FINALv3211122.pdf (accessed 23 May 2025). The plans were revised in 2024: Environment Agency, *A Summary of England's Revised Draft Regional and Water Resources Management Plans*.

75 Environment Agency, *A Summary of England's Revised Draft Regional and Water Resources Management Plans* (2024).

76 Environment Agency, 'Appendix A: smart metering in revised draft water resources management plans' (2024), https://www.gov.uk/government/publications/a-review-of-englands-draft-regional-and-water-resources-management-plans/appendix-a-smart-metering-in-draft-water-resources-management-plans (accessed 23 May 2025).

77 Environment Agency, 'Appendix A: smart metering in revised draft water resources management plans' (2024).

78 Artesia Consulting, *The Long Term Potential for Deep Reductions in Household Water Demand*, report for Ofwat (2018), https://www.ofwat.gov.uk/wp-content/uploads/2018/05/The-long-term-potential-for-deep-reductions-in-household-water-demand-report-by-Artesia-Consulting.pdf (accessed 23 May 2025).

79 Ofwat, 'Leakage Dataset – March 2023', 22 March 2023, https://www.ofwat.gov.uk/publication/leakage-dataset-march-2023/ (accessed 23 May 2025).

80 Environment Agency, *A Summary of England's Revised Draft Regional and Water Resources Management Plans*, section 4.2.

81 Source: Defra: 'E8 Efficient use of water', tables E8a and E8b, https://oifdata.defra.gov.uk/themes/natural-resources/E8/ (accessed 23 May 2025); Notes: weighted average. Data relates to April–March (financial year) and give three-year moving averages. This aligns with Ofwat targets/reporting and helps to reduce sensitivity to events such as weather conditions. Water leaks data is converted from megalitres to per person using population data from Nomis.

82 Smart Meter Statistics in Great Britain, 'Smart meter statistics in Great Britain: quarterly report to end September 2024', 28 November 2024, https://assets.publishing.service.gov.uk/media/6745aec3da210676b4ffe117/Q3_2024_Smart_Meters_Statistics_Report.pdf (accessed 23 May 2025).

83 National Infrastructure Commission, *Preparing for a Drier Future*, p. 10.

Notes

84 Claudio Daminato, Eugenio Diaz-Farina, Massimo Filippini and Noemi Padrón-Fumero, 'The impact of smart meters on residential water consumption: evidence from a natural experiment in the Canary Islands', *Resource and Energy Economics*, 64 (2021), 101221.

85 Environment Agency, *A Summary of England's Revised Draft Regional and Water Resources Management Plans*, section 4.2.

86 Government Office for Science, 'Future of the subsurface: urban water management in the UK (annex)', 18 November 2024, https://www.gov.uk/government/publications/future-of-the-subsurface-report/future-of-the-subsurface-urban-water-management-in-the-uk-annex#fn (accessed 23 May 2025).

87 Kevin Rawlinson, 'Dual-flush toilets "wasting more water than they save"', *The Guardian*, 29 September 2020, https://www.theguardian.com/environment/2020/sep/29/dual-flush-toilets-wasting-more-water-than-they-save (accessed 23 May 2025).

88 Environment Agency, *National Flood and Coastal Erosion Risk Management Strategy for England* (2024), https://assets.publishing.service.gov.uk/media/5f6b6da6e90e076c182d508d/023_15482_Environment_agency_digitalAW_Strategy.pdf (accessed 23 May 2025).

89 For an earlier, less alarming view, see National Infrastructure Commission, *Reducing the Risk of Surface Water Flooding*.

90 Environment Agency, *National Flood and Coastal Erosion Risk Management Strategy for England*, p. 4.

91 Environment Agency, *National Flood and Coastal Erosion Risk Management Strategy for England*, p. 5.

92 Environment Agency, *National Flood and Coastal Erosion Risk Management Strategy for England*, p. 5.

93 Defra, *Our Integrated Plan for Delivering Clean and Plentiful Water* (2023), https://assets.publishing.service.gov.uk/media/6492c8b35f7bb7000c7fae61/plan_for_water.pdf (accessed 23 May 2025).

94 Defra, *Our Integrated Plan for Delivering Clean and Plentiful Water*, p. 7.

95 Defra, *Our Integrated Plan for Delivering Clean and Plentiful Water*, p. 5.

96 Defra, *Our Integrated Plan for Delivering Clean and Plentiful Water*.

97 Susdrain (undated), 'Sustainable drainage', https://www.susdrain.org/delivering-suds/using-suds/background/sustainable-drainage.html (accessed 23 May 2025).

98 UK Government, 'Trees and woodlands provide over £400m each year in fight against flooding, new study finds', 13 January 2023, https://

Notes

www.gov.uk/government/news/trees-and-woodlands-provide-over-400m-each-year-in-fight-against-flooding-new-study-finds (accessed 23 May 2025).
99 Defra, *Our Integrated Plan for Delivering Clean and Plentiful Water*, p. 8.
100 Defra, *Our Integrated Plan for Delivering Clean and Plentiful Water*, p. 9.

Chapter 4

1 Ruth Sylvester, P. Hutchings and A. Mdee, 'Defining and acting on water poverty in England and Wales', *Water Policy*, 25.5 (2023), 505.
2 UN Environment Programme, 'GOAL 6: Clean water and sanitation', *Sustainable Development Goals* (2024), https://www.unep.org/topics/sustainable-development-goals/why-do-sustainable-development-goals-matter/goal-6 (accessed 23 May 2025).
3 Desmond McKibbin, 'Extent of the mains water network in England, Wales, Scotland and the Republic of Ireland', Research Paper NIAR 183/10, *Northern Ireland Assembly* (2010), https://archive.niassembly.gov.uk/regional/2007mandate/research/pdf/Extent_of_the_Mains_Water.pdf (accessed 23 May 2025).
4 Sylvester et al., 'Defining and acting on water poverty', p. 495.
5 Citizens Advice, 'WaterSure scheme – help with paying water bills', 20 February 2020, https://www.citizensadvice.org.uk/consumer/water/problems-with-paying-your-water-bill/watersure-scheme-help-with-paying-water-bills/ (accessed 23 May 2025).
6 Defra, 'Company Social Tariffs: Guidance to water and sewerage undertakers and the Water Services Regulation Authority under Section 44 of the Flood and Water Management Act 2010' (2012), https://assets.publishing.service.gov.uk/media/5a798ef740f0b63d72fc6c46/pb13787-social-tariffs-guidance.pdf (accessed 23 May 2025).
7 CEPA, *Quantitative Analysis of Water Poverty in England and Wales*, report for Water UK (2021), p. 9. https://www.water.org.uk/sites/default/files/wp/2021/04/Quantitative-analysis-of-water-poverty-in-England-and-Wales.pdf (accessed 23 May 2025).
8 Ofwat, 'Cost of living: wave two: Water customers' experiences', 1 December 2022, https://www.ofwat.gov.uk/publication/cost-of-

Notes

living-wave-two-water-customers-experiences/ (accessed 23 May 2025).

9 Ofwat, *Analysis of Household Customer Debt* (2025), p. 7, https://www.ofwat.gov.uk/wp-content/uploads/2025/01/Analysis-of-household-customer-debt.pdf (accessed 23 May 2025).

10 CCW, '2 in 5 households say they will find it difficult to afford water bill increases', press release, 6 November 2024, https://www.ccw.org.uk/news/2-in-5-households-say-they-will-find-it-difficult-to-afford-water-bill-increases/ (accessed 23 May 2025).

11 Defra, 'Changes to WaterSure as a result of the introduction of Universal Credit' (2012), https://assets.publishing.service.gov.uk/government/uploads/system/uploads/attachment_data/file/82677/watersure-condoc-121018.pdf (accessed 23 May 2025).

12 CCW, 'Call for input – improving the WaterSure financial support scheme', 17 July 2024, https://www.ccw.org.uk/publication/call-for-input-improving-the-watersure-financial-support-scheme/ (accessed 23 May 2025). Note: the CCW is an independent non-departmental public body (sponsored by Defra) that supports and advocates for water customers.

13 CCW (undated), 'Affordability Review: outcomes', https://www.ccw.org.uk/our-work/affordability-and-vulnerability/affordability-review/affordability-review-outcomes/ (accessed 23 May 2025). Note that there are 17 regulated companies providing water only or water and sewerage services. The ten large water and sewerage companies are the focus of the analysis presented in this book.

14 House of Lords Industry and Regulators Committee, *The Affluent and the Effluent: Cleaning Up Failures in Water and Sewage Regulation*, 2022–23 HL Paper 166 (2023), paras 125 and 127, https://publications.parliament.uk/pa/ld5803/ldselect/ldindreg/166/166.pdf (accessed 23 May 2025).

15 Ruby Flanagan, 'Martin Lewis' MSE issues warning as millions of Brits missing out on £160 bill discount', *Mirror*, 11 April 2024, https://www.mirror.co.uk/money/martin-lewis-mse-issues-warning-32564224 (accessed 23 May 2025).

16 Water UK, 'Hundreds of thousands more people to receive help paying their water bills', press release, 19 January 2023, https://www.water.org.uk/news-views-publications/news/

Notes

hundreds-thousands-more-people-receive-help-paying-their-water-bills (accessed 23 May 2025).

17 Sylvester et al., 'Defining and acting on water poverty', p. 500.
18 Water UK, 'Water UK response to call for evidence for the independent review of affordability support for financially vulnerable water customers in England and Wales', 15 December 2020, p. 1, https://www.water.org.uk/sites/default/files/wp/2020/12/Water-UK-response-to-call-for-evidence-for-CCW-Affordability-Review.pdf (accessed 23 May 2025).
19 Water UK, 'Water companies propose largest ever investment', press release, 2 October 2023, https://www.water.org.uk/news-views-publications/news/water-companies-propose-largest-ever-investment (accessed 23 May 2025).
20 Water UK (undated), 'Worried about your water bill? Help is ready for those who need it', https://www.water.org.uk/customers/help-bills (accessed 16 December 2024).
21 Ofwat, *Our Final Determinations for the 2024 Price Review: Sector Summary* (2024), p. 20, https://www.ofwat.gov.uk/wp-content/uploads/2024/12/PR24-FD-sector-summary.pdf (accessed 23 May 2025).
22 Defra, 'Plans to help households with water bills', press release, 5 April 2011, https://www.gov.uk/government/news/plans-to-help-households-with-water-bills (accessed 23 May 2025).
23 CCW, 'Call for input – improving the WaterSure financial support scheme'.
24 Sylvester et al., 'Defining and acting on water poverty', p. 501.
25 Ofwat, *Our Final Determinations for the 2024 Price Review: Sector Summary*, p. 21.
26 Gill Plimmer and Jim Pickard, 'Higher England and Wales water bills risk "crisis" for households', *Financial Times*, 12 July 2023, https://www.ft.com/content/fb27dc56-73cf-4765-81a5-8a2aa6eae055 (accessed 23 May 2025).
27 CCW, *Water Worries. Affordability Research 2025* (2025), p. 16, https://www.ccw.org.uk/publication/water-worries-affordability-research-2025/ (accessed 23 May 2025); Independent Age, 'Single water social tariff: open letter', 19 August 2024, https://www.independentage.org/single-water-social-tariff-open-letter (accessed 23 May 2025); Social Market Foundation, 'The shape of a social

Notes

tariff', briefing, 16 January 2025, https://www.smf.co.uk/publications/unified-social-tariff-water/ (accessed 23 May 2025).

28 Jeevan Jones, 'Water companies to triple customer support, but further reforms needed', Water UK, 30 January 2025, https://www.water.org.uk/news-views-publications/views/new-affordability-plans-blog (accessed 23 May 2025).

29 Citizens Information (undated), 'Household water charge for excess use', https://www.citizensinformation.ie/en/housing/water-and-coasts/water-charges/ (accessed 23 May 2025).

30 Environment Agency, 'Appendix A: Smart metering in revised draft water resources management plans', 20 December 2024, https://www.gov.uk/government/publications/a-review-of-englands-draft-regional-and-water-resources-management-plans/appendix-a-smart-metering-in-draft-water-resources-management-plans (accessed 23 May 2025). Note: data refers to England.

31 Andrew Capstick, 'Cut your water bills: Big meter savings, freebies & more', *MoneySavingExpert*, 13 December 2024, https://www.moneysavingexpert.com/utilities/cut-water-bills/ (accessed 23 May 2025).

32 This is the weighted mean average household water and sewerage bill of all households in England and Wales in 2023. This is the last full year of information at the time of writing.

33 Source: ONS, Family spending workbook 5: expenditure on housing, dataset, 23 August 2024, https://www.ons.gov.uk/peoplepopulationandcommunity/personalandhouseholdfinances/expenditure/datasets/familyspendingworkbook5expenditureonhousing (accessed 23 May 2025). Note: water charges annualised from weekly spend. The data is from household spending survey collected between April and March 2023. Hence, it does not directly align to company revenue because of timing differences and different accounting periods, and the regions do not directly align with the water company territories.

34 Discover Water (undated), 'Average annual water and sewerage charges across England and Wales households', http://www.discoverwater.co.uk/annual-bill (accessed 16 December 2024).

35 Water UK (undated), '£108bn investment proposed for the biggest upgrade of our water and sewage systems', https://www.water.org.uk/investing-future/pr24 (accessed 23 May 2025).

Notes

36 National Audit Office, *Infrastructure Investment: The Impact on Consumer Bills*, 2013–14 HC 812-I (2013), https://www.nao.org.uk/reports/infrastructure-investment-impact-consumer-bills-2/ (accessed 23 May 2025).

37 ONS, 'Living Costs and Food Survey: user guidance and technical information for the Living Costs and Food Survey', 16 February 2017, https://www.ons.gov.uk/peoplepopulationandcommunity/personalandhouseholdfinances/incomeandwealth/methodologies/livingcostsandfoodsurvey (accessed 23 May 2025).

38 Source: Adapted from ONS, Family Spending 2023, table 2.3 (workbook 5: expenditure on housing), https://www.ons.gov.uk/peoplepopulationandcommunity/personalandhouseholdfinances/expenditure/datasets/familyspendingworkbook5expenditureonhousing (accessed 23 May 2025), various years and company reports and accounts. Notes: the totals include all categories documented in the LCF, including those not part of the 'COICOP' total expenditure. The calculations are based on data from table 2.3 in Workbook 5, adjusted to correspond with the annual revenue of water companies and refers solely to households in England and Wales.

39 Source: Adapted from Family Spending 2023 (workbook 5: expenditure on housing). Note: data refers to financial year April 2022–March 2023.

40 Ofwat, *Our Final Determinations for the 2024 Price Review: Sector Summary*, p. 3.

41 Gill Plimmer, 'Thames Water customers will have paid £540mn for London's "Super Sewer"', *Financial Times*, 26 April 2024, https://www.ft.com/content/1b58a0f7-cf7a-4d39-9ebb-93db7172ece3 (accessed 23 May 2025).

42 Source: Adapted using the ONS, Family Spending 2023, table 2.3 (workbook 5 : expenditure on housing), https://www.ons.gov.uk/peoplepopulationandcommunity/personalandhouseholdfinances/expenditure/datasets/familyspendingworkbook5expenditureonhousing (accessed 23 May 2025). Notes: the totals include all categories documented in the LCF, including those not part of the 'COICOP' total expenditure. The calculations are based on data from table 2.3 in Workbook 5, adjusted to correspond with the annual revenue of water companies, and refers solely to households in England and Wales.

Notes

43 See, for example, the responses in December 2024 to the final determination by Ofwat, which includes how much customer charges will be allowed to increase by 2030, as captured in Robert Smith, Kieran Smith, Jim Pickard and Euan Healy, 'Why UK water regulator's decision on bills is set to stoke turmoil', *Financial Times*, 19 December 2024, https://www.ft.com/content/921610e9-0604-4ab4-9ac3-cc8ee9292789 (accessed 23 May 2025).

44 Philipp Lausberg and Tijn Croon, 'Europe must fight energy poverty more effectively', *European Policy Centre*, 19 January 2023, https://www.epc.eu/publication/-Europe-must-fight-energy-poverty-more-effectively-4da8dc/ (accessed 23 May 2025).

45 For example, Age UK estimated in 2024 that more than one-third of pensioner households (800,000) in the UK that are eligible for Pension Credit do not claim, and that 21% (310,000) of eligible households do not claim Housing Benefit. Age UK, 'Briefing. Benefit take-up and older people' (2024), https://www.ageuk.org.uk/siteassets/documents/reports-and-publications/reports-and-briefings/money-matters/benefit-take-up-briefing-may-2024-.pdf (accessed 23 May 2025).

46 Ofwat notes that companies have a 'plan' for numbers of customers who will be supported, 'not a target'. Ofwat, *Summary of Companies' Published Plans for Affordability for 2025–30*, PR24 Final Determinations (2024), p. 9, https://www.ofwat.gov.uk/wp-content/uploads/2024/12/Summary-of-water-companies-published-plans-for-affordability-for-2025-30-republished-19-December-2024.pdf (accessed 23 May 2025).

47 CCW, *Water Worries. Affordability Research 2025*, p. 21.

48 Ofwat, *Summary of Companies' Published Plans for Affordability for 2025–30*, p. 10.

49 Ofwat, *Transcript of Investor Briefing*, PR24 Final Determinations (2024), p. 2, https://www.ofwat.gov.uk/wp-content/uploads/2024/12/PR24-final-determinations-City-briefing-transcript.pdf (accessed 23 May 2025).

50 Ofwat, *Summary of Companies' Published Plans for Affordability for 2025–30*, p. 9.

51 Source: Adapted from Family Spending (workbook 5: expenditure on housing), ONS, https://www.ons.gov.uk/peoplepopulationandcommunity/personalandhouseholdfinances/expenditure/datasets/

Notes

familyspendingworkbook5expenditureonhousing (accessed 23 May 2025), Ofwat and water company data.

52 Source: Authors' calculations based on adaptation from Family Spending (workbook 5: expenditure on housing), ONS, https://www.ons.gov.uk/peoplepopulationandcommunity/personalandhouseholdfinances/expenditure/datasets/familyspendingworkbook5expenditureonhousing (accessed 23 May 2025), Ofwat and water company data. Notes: the totals include all categories documented in the LCF, including those not part of the ''COICOP' total expenditure. The calculations are based on data from table 2.3 in Workbook 5, adjusted to correspond with the annual revenue of water companies and refers solely to households in England and Wales.

53 Source: Authors' calculations based on adaptation from Family Spending (workbook 5: expenditure on housing), ONS, https://www.ons.gov.uk/peoplepopulationandcommunity/personalandhouseholdfinances/expenditure/datasets/familyspendingworkbook5expenditureonhousing (accessed 23 May 2025), Ofwat and water company data. Notes: the totals include all categories documented in the LCF, including those not part of the 'COICOP' total expenditure. The calculations are based on data from table 2.3 in Workbook 5, adjusted to correspond with the annual revenue of water companies and refers solely to households in England and Wales.

54 See Exhibit 4.8 for this simulation and a comparison with the actual system and the flat rate simulation.

55 Department of Work and Pensions, 'Universal Credit statistics, 29 April 2013 to 10 October 2024', 28 November 2024, https://www.gov.uk/government/statistics/universal-credit-statistics-29-april-2013-to-10-october-2024/universal-credit-statistics-29-april-2013-to-10-october-2024 (accessed 23 May 2025).

56 HM Revenue & Customs, 'Child benefit statistics: annual release, data at August 2023', 17 April 2024, https://www.gov.uk/government/statistics/child-benefit-statistics-annual-release-august-2023/child-benefit-statistics-annual-release-data-at-august-2023#take-up-rate-of-child-benefit (accessed 23 May 2025).

57 House of Commons Library, 'Student loan statistics', UK Parliament, 5 December 2024, https://commonslibrary.parliament.uk/research-briefings/sn01079/ (accessed 23 May 2025).

Notes

58 HM Revenue & Customs, 'Fascinating facts about Self-Assessment', 18 January 2021, https://www.gov.uk/government/news/fascinating-facts-about-self-assessment (accessed 23 May 2025).
59 Craig Lowrey and Kate Mulvany, 'How effective were the UK's energy bill support schemes?', Cornwall Insight, 28 March 2024, https://www.cornwall-insight.com/thought-leadership/blog/how-effective-were-the-uks-energy-bill-support-schemes/ (accessed 23 May 2025).
60 Suzanna Hinson and Paul Bolton, *Fuel Poverty in the UK*, House of Commons Library Research Briefing Paper CBP 8730, 14 April 2025, https://researchbriefings.files.parliament.uk/documents/CBP-8730/CBP-8730.pdf (accessed 23 May 2025).
61 Brigid Francis-Devine, Xameerah Malik and Nerys Roberts, *Food Poverty: Households, Food Banks and Free School Meals*, House of Commons Library Research Briefing Paper CBP 9209, 2 September 2024, https://researchbriefings.files.parliament.uk/documents/CBP-9209/CBP-9209.pdf (accessed 23 May 2025).
62 Source: Authors' calculations based on adaptation from Family Spending 2023 (workbook 5: expenditure on housing), ONS, https://www.ons.gov.uk/peoplepopulationandcommunity/personalandhouseholdfinances/expenditure/datasets/familyspendingworkbook5expenditureonhousing (accessed 23 May 2025), Ofwat and water company data.
63 Robert Smith and Gill Plimmer, 'Thames Water creditors head for courtroom showdown over £3bn emergency loan', *Financial Times*, 12 December 2024, https://www.ft.com/content/2b77b43a-1ba8-4107-a104-8f6a4c8e334e (accessed 23 May 2025).

Chapter 5

1 Mark Abrams, Richard Rose and Rita Hinden, *Must Labour Lose?* (Penguin, 1960).
2 Joe Moran, 'Mass observation, market research and the birth of the focus group 1937–97', *Journal of British Studies*, 47 (2008), 827–51; see also Andrew Marr, 'How Blair put the media in a spin', BBC News, 10 May 2007, http://news.bbc.co.uk/1/hi/uk_politics/6638231.stm (accessed 23 May 2025).

Notes

3 Philip Gould, *The Unfinished Revolution* (Abacus, 2011), p. 328.
4 Gould, *The Unfinished Revolution*, p. 327.
5 House of Commons Environmental Audit Committee, *Water Quality in Rivers*, 2021–22 HC 74 (2022), p. 5, https://publications.parliament.uk/pa/cm5802/cmselect/cmenvaud/74/report.html (accessed 23 May 2025).
6 House of Lords Industry and Regulators Committee, *The Affluent and the Effluent: Cleaning up Failures in Water and Sewage Regulation*, 2022–23 HL Paper 166 (2023), pp. 4–5, https://committees.parliament.uk/publications/34458/documents/189872/default/ (accessed 23 May 2025). A follow-up inquiry was also conducted in 2023. The full chronology is here: https://lordslibrary.parliament.uk/cleaning-up-failures-in-water-and-sewage-regulation-industry-and-regulators-committee-report/ (accessed 23 May 2025).
7 https://webarchive.nationalarchives.gov.uk/ukgwa/20250327094044/https:/nic.org.uk/ (accessed 23 May 2025).
8 National Infrastructure Commission, *Second National Infrastructure Assessment* (2023), ch. 4, https://webarchive.nationalarchives.gov.uk/ukgwa/20250327100337/https:/nic.org.uk/studies-reports/national-infrastructure-assessment/second-nia/ (accessed 23 May 2025).
9 Office for Environmental Protection, https://www.theoep.org.uk (accessed 23 May 2025).
10 Office for Environmental Protection, *Progress in Improving the Natural Environment in England 2022/2023* (2024), p. 10, https://www.theoep.org.uk/report/government-remains-largely-track-meet-its-environmental-ambitions-finds-oep-annual-progress (accessed 23 May 2025).
11 Office for Environmental Protection, *Progress in Improving the Natural Environment in England 2022/2023*, p. 11.
12 Office for Environmental Protection, *Progress in Improving the Natural Environment in England 2022/2023*, press release.
13 Green Party, 'Our manifesto – bringing nature back to life' (2024), https://greenparty.org.uk/about/our-manifesto/bringing-nature-back-to-life/ (accessed 23 May 2025).
14 Reform UK, 'Our contract with you' (2024), p. 17, https://assets.nationbuilder.com/reformuk/pages/253/attachments/original/1718625371/Reform_UK_Our_Contract_with_You.pdf?1718625371 (accessed 23 May 2025).

Notes

15 Liberal Democrats, 'For a fair deal. Manifesto 2024' (2024), section 12, https://www.libdems.org.uk/manifesto (accessed 23 May 2025).

16 Gill Plimmer and Jim Pickard, 'Labour plans new water regulator for England and Wales', *Financial Times*, 5 May 2023, https://www.ft.com/content/602a009b-6a41-4528-9804-22f7a5080cc3 (accessed 23 May 2025).

17 Labour Party, 'Our plan to change Britain' (2024), p. 59, https://labour.org.uk/change/ (accessed 23 May 2025).

18 Liberal Democrats, 'For a fair deal. Manifesto 2024', section 12.

19 Labour Party, 'Clear water. Labour's vision for a modern and transparent publicly-owned water system' (2018), pp. 4–5, https://www.labour.org.uk/wp-content/uploads/2018/09/Conference-2018-Water-pamphlet-FINAL.pdf (accessed 23 May 2025).

20 Conservatives, 'The Conservative and Unionist Party Manifesto 2024' (2024), pp. 66–7, https://public.conservatives.com/static/documents/GE2024/Conservative-Manifesto-GE2024.pdf (accessed 23 May 2025).

21 Liberal Democrats, 'For a fair deal. Manifesto 2024', Foreword.

22 Labour Party, 'Our plan to change Britain', p. 59.

23 Steve Reed, 'We'll charge water bosses and ban their bonuses until they fix this toxic mess', *Mail Online*, 27 July 2024, https://www.dailymail.co.uk/news/article-13679409/steve-reed-environment-secretary-water-sewage.html?ito=email_share_article-top (accessed 23 May 2025).

24 Defra, 'Government introduces new bill to clean up water sector', press release, 18 July 2024, https://deframedia.blog.gov.uk/2024/07/18/government-introduces-new-bill-to-clean-up-water-sector/ (accessed 23 May 2025).

25 Defra, 'Government introduces new bill to clean up water sector'.

26 Defra, 'Government introduces new bill to clean up water sector'.

27 Defra, 'Government introduces new bill to clean up water sector'.

28 Defra, 'Steve Reed speech on the Water (Special Measures) Bill', 5 September 2024, https://www.gov.uk/government/speeches/steve-reed-speech-on-the-water-special-measures-bill (accessed 23 May 2025).

29 Clive Mottram, 'New water industry Special Administration Regime', Weightmans, 24 April 2024, https://www.weightmans.com/

Notes

media-centre/news/new-water-industry-special-administration-regime/ (accessed 23 May 2025).

30 Michael Savage, 'Public anger over water bill rises is justified, says UK environment secretary', *The Observer*, 15 December 2024, https://www.theguardian.com/money/2024/dec/15/water-bill-rises-england-wales-environment-secretary (accessed 23 May 2025).

31 Defra and Welsh Government, 'Governments launch largest review of sector since privatisation', press release, 22 October 2024, https://www.gov.uk/government/news/governments-launch-largest-review-of-sector-since-privatisation (accessed 23 May 2025).

32 Lucie Heath, 'Labour has no real plan to fix UK rivers, Feargal Sharkey warns', *The i Paper*, 21 August 2024, https://inews.co.uk/news/labour-no-real-plan-fix-uk-rivers-3237514 (accessed 23 May 2025).

33 Jonah Fisher, 'Anti-pollution law to threaten water bosses with jail', BBC News, 4 Sept 2024, https://www.bbc.co.uk/news/articles/cwy5dwlwkwro? (accessed 23 May 2025).

34 Defra, 'February 2022: the government's strategic priorities for Ofwat', policy paper, 28 March 2022, https://www.gov.uk/government/publications/strategic-policy-statement-to-ofwat-incorporating-social-and-environmental-guidance/february-2022-the-governments-strategic-priorities-for-ofwat (accessed 23 May 2025).

35 R. Hern, 'Competition and access pricing in the UK water industry', *Utilities Policy*, 10.3–4 (2001), 117–27.

36 Dieter Helm, 'Thirty years after water privatization – is the English model the envy of the world?', *Oxford Review of Economic Policy*, 36.1 (2020), 82.

37 Competition and Markets Authority, 'Anglian Water Services Limited, Bristol Water plc, Northumbrian Water Limited and Yorkshire Water Services Limited Price Determinations', 17 March 2021, https://assets.publishing.service.gov.uk/media/604fa141e90e077fe7a5f45a/-_CMA_water_redeterminations_-_summary_-_online_version_—_-.pdf (accessed 23 May 2025).

38 Water Magazine, 'Six companies to appeal Ofwat's PR24 Final Determination', 19 February 2025, https://www.watermagazine.co.uk/2025/02/19/six-water-companies-to-appeal-ofwats-pr24-final-determination/ (accessed 23 May 2025).

Notes

39 Ofwat, 'What the 2024 price review means for water customers' (2024), https://www.ofwat.gov.uk/regulated-companies/price-review/2024-price-review/what-it-means-for-customers-and-water-bills/#UnitedUtil (accessed 23 May 2025).

40 Gill Plimmer, 'Under-fire UK water companies lash out at "labyrinthine" regulatory system', *Financial Times*, 20 May 2024, https://www.ft.com/content/856b49ee-5c89-44ad-96bb-0d4ffff4fcab (accessed 23 May 2025).

41 Ofwat, 'Written evidence submitted by Ofwat to the Environment, Food and Rural Affairs committee inquiry into the regulation of the water industry', House of Commons, 6 June 2018, para. 1, https://committees.parliament.uk/writtenevidence/91274/pdf/ (accessed 23 May 2025).

42 House of Lords, 'Industry and Regulators Committee. Corrected oral evidence: the work of Ofwat', 5 July 2022, Q44, https://committees.parliament.uk/oralevidence/10534/html/ (accessed 23 May 2025).

43 Ofwat, *Gender and Ethnicity Pay Gap Report 2023* (2024), https://www.ofwat.gov.uk/wp-content/uploads/2024/03/Gender-and-ethnicity-pay-gap-report-2023.pdf (accessed 23 May 2025).

44 Ofwat, 'Written evidence submitted by Ofwat to the Environment, Food and Rural Affairs committee inquiry into the regulation of the water industry', House of Commons, 6 June 2018, para. 5, https://committees.parliament.uk/writtenevidence/91274/pdf/ (accessed 23 May 2025).

45 Defra, 'The government's strategic priorities for Ofwat', 28 March 2022, https://www.gov.uk/government/publications/strategic-policy-statement-to-ofwat-incorporating-social-and-environmental-guidance/february-2022-the-governments-strategic-priorities-for-ofwat (accessed 23 May 2025).

46 Ian Byatt, 'The regulation of water services in the UK', *Utilities Policy*, 24 (2012), 6.

47 Byatt, 'The regulation of water services in the UK', p. 7.

48 Julie Froud and Karel Williams, 'Private equity and the culture of value extraction', *New Political Economy*, 12.3 (2007), 405–20.

49 Julie Froud, Sukhdev Johal, Adam Leaver and Karel Williams, *Financialisation and Strategy* (Routledge, 2006).

Notes

50 Alex Lawson and Sandra Laville, 'Ofwat investigates whether Thames Water dividend is licensing breach', *The Guardian*, 5 December 2023, https://www.theguardian.com/business/2023/dec/05/ofwat-investigates-whether-thames-water-dividend-is-licensing-breach (accessed 23 May 2025).

51 Gill Plimmer, Thames water to be investigated over financial stability and dividends', *Financial Times*, 5 December 2023, https://www.ft.com/content/d6438dac-847a-4842-aa07-637ae049f141 (accessed 23 May 2025).

52 Source: Company annual report and accounts, 2022/23.

53 Centrus, 'Centrus advises Yorkshire water on restructuring index linked derivatives with mandatory breaks in 2023', 19 April 2021, https://centrusfinancial.com/credential/centrus-advises-yorkshire-water-on-restructuring-index-linked-derivatives-with-mandatory-breaks-in-2023/ (accessed 23 May 2025).

54 Carmen Aguilar García, Anna Leach and Sandra Laville, 'How we calculated the proportion of revenue English water firms use to pay off debt', *The Guardian*, 18 December 2023, https://amp.theguardian.com/business/2023/dec/18/how-calculated-revenue-english-water-firms-pay-debt (accessed 23 May 2025).

55 Kate Bayliss and David Hall, 'Bringing water into public ownership: costs and benefits' (2017), https://gala.gre.ac.uk/id/eprint/17277/3/17277%20HALL_Bringing_Water_into_Public_Ownership_2017.pdf (accessed 23 May 2025).

56 Ofwat, *Putting the Sector in Balance: Position Statement on PR19 Business Plans* (2018), p. 37, https://www.ofwat.gov.uk/wp-content/uploads/2018/04/Putting-the-sector-in-balance-position-statement-on-PR19-business-plans-FINAL2.pdf (accessed 23 May 2025).

57 Ofwat, *Putting the Sector in Balance*, pp. 8–9.

58 Ofwat, *Putting the Sector in Balance*, pp. 5–6.

59 House of Lords Industry and Regulators Committee, 'Corrected oral evidence: The work of Ofwat', 25 October 2022, Q123, https://committees.parliament.uk/oralevidence/11401/html/ (accessed 23 May 2025).

60 Euan Healy and Gill Plimmer, 'Southern water sinks to record operating loss', *Financial Times*, 28 November 2024, https://www.ft.com/content/2e070d6e-a7bc-4126-b65b-ab2d89db27aa (accessed 23

Notes

May 2025); Robert Smith and Gill Plimmer, 'Moody's downgrade pushed Southern Water closer to default', *Financial Times*, 14 November 2024, https://www.ft.com/content/8b56376d-7c1b-4aa9-bf9e-63e6a45c3ede (accessed 23 May 2025).

61 'About us – Environment Agency', https://www.gov.uk/government/organisations/environment-agency/about (accessed 23 May 2025).

62 Ofwat, *Gender and Ethnicity Pay Gap Report 2023*.

63 Defra, *Our Integrated Plan for Delivering Clean and Plentiful Water* (2023), p. 16, https://assets.publishing.service.gov.uk/media/6492c8b35f7bb7000c7fae61/plan_for_water.pdf (accessed 23 May 2025).

64 Defra, 'The government's strategic priorities for Ofwat', pp. 3–4.

65 Defra (undated), 'About the Defra in the media blog', https://deframedia.blog.gov.uk/about-the-defra-in-the-media-blog/ (accessed 23 May 2025).

66 Esme Stallard, Becky Dale, Jonah Fisher and Sophie Woodcock, 'Water firms illegally spilled sewage on dry days – data suggests', BBC News, 5 September 2023, https://www.bbc.co.uk/news/science-environment-66670132 (accessed 23 May 2025).

67 Environment Agency, 'Water and sewerage companies in England: EPA metric guide for 2022', 12 July 2023, https://www.gov.uk/government/publications/water-and-sewerage-companies-in-england-environmental-performance-report-2022/water-and-sewerage-companies-in-england-epa-metric-guide-for-2022 (accessed 23 May 2025).

68 Environment Agency, 'Water and sewerage companies in England: EPA metric guide for 2022'; Environment Agency, 'Tougher regulation as data shows water companies underperforming', press release, 23 July 2024, https://www.gov.uk/government/news/tougher-regulation-as-data-shows-water-companies-underperforming (accessed 23 May 2025).

69 Joe Crowley, 'Raw sewerage "cover-up" at Windermere World Heritage Site', BBC News, 4 December 2023, https://www.bbc.co.uk/news/science-environment-67567323 (accessed 23 May 2025).

70 Environment Agency, 'Water and sewerage companies in England: EPA metric guide for 2022', section 4.3.

71 BBC, 'The water pollution cover up', *Panorama*, 4 December 2023.

Notes

72 Water UK, 'English water companies publish world-leading plan to remove 4 million spills from our rivers and coasts', 12 March 2024, https://www.water.org.uk/news-views-publications/news/english-water-companies-publish-world-leading-plan-remove-4-million (accessed 23 May 2025).

73 Water UK, *National Storm Overflows Plan for England* (2024), p. 25, https://www.water.org.uk/sites/default/files/2024-03/WEB_Water%20UK%20National%20Storm%20Overflows%20Plan%20for%20England_0.pdf (accessed 23 May 2025).

74 UK Parliament, 'Rivers: sewage', question for Defra UIN 10064, 21 October 2024, https://questions-statements.parliament.uk/written-questions/detail/2024-10-21/10064 (accessed 23 May 2025).

75 House of Commons Environmental Audit Committee, *Oral Evidence: the Environmental Protection Work of the Environment Agency*, HC 702 2023–24, 24 April 2024, Q.15, https://committees.parliament.uk/oralevidence/14718/pdf/ (accessed 23 May 2025).

76 House of Commons Environment, Food and Rural Affairs Committee, *Regulation of the Water Industry*, HC 1041, 8th Report 2017-19 (2018), p. 7, https://publications.parliament.uk/pa/cm201719/cmselect/cmenvfru/1041/1041.pdf (accessed 23 May 2025).

77 Environment Agency, 'Changes to Environment Agency's abstraction charges to safeguard water supplies for people and wildlife', press release, 9 March 2022, https://www.gov.uk/government/news/changes-to-environment-agencys-abstraction-charges-to-safeguard-water-supplies-for-people-and-wildlife (accessed 23 May 2025).

78 Environment Agency, *The State of the Environment: Water Resources* (2018), p. 1, https://assets.publishing.service.gov.uk/media/60fad187d3bf7f044fbb0763/State_of_the_environment_water_resources_report.pdf (accessed 23 May 2025).

79 House of Commons Environmental Audit Committee, *Oral Evidence: the Environmental Protection Work of the Environment Agency*.

80 BBC News, 'UK floods: "complete rethink needed" on flood defences', 28 December 2015, https://www.bbc.co.uk/news/uk-35188146 (accessed 23 May 2025).

81 Environment Agency, 'Tougher regulation as data shows water companies underperforming'.

Notes

82 Environment Agency, 'Record £90m fine for Southern Water following EA prosecution', press release, 9 July 2021, https://www.gov.uk/government/news/record-90m-fine-for-southern-water-following-ea-prosecution (accessed 23 May 2025).

83 Ofwat, 'Thames, Yorkshire and Northumbrian water face £168 million penalty following sewage investigation', press release, 6 August 2024, https://www.ofwat.gov.uk/thames-yorkshire-and-northumbrian-water-face-168-million-penalty-following-sewage-investigation/ (accessed 23 May 2025).

84 John Collingridge and Anna Isaac, 'Thames Water asks Ofwat for leniency on costs and fines as it tries to attract buyers', *The Guardian*, 15 March 2025, https://www.theguardian.com/business/2025/mar/14/thames-water-asks-ofwat-to-be-spared-fines-costs (accessed 23 May 2025).

85 Defra, 'Unlimited penalties introduced for those who pollute environment', press release, 11 December 2023, https://www.gov.uk/government/news/unlimited-penalties-introduced-for-those-who-pollute-environment (accessed 23 May 2025).

86 National Audit Office, *Understanding Storm Overflows: Exploratory Analysis of Environment Agency Data* (2021), p. 7, https://www.nao.org.uk/wp-content/uploads/2021/09/010856-001-Understanding-storm-overflows-FINAL2-accessible.pdf (accessed 23 May 2025); Defra, 'Environment Agency response to Panorama investigation', 4 December 2023, https://deframedia.blog.gov.uk/2023/12/04/environment-agency-response-to-panorama-investigation/ (accessed 23 May 2025). This blog post notes that monitoring of storm overflows was at 10% in 2015.

87 Defra, 'Storm overflows monitoring hits 100% target', press release, 30 December 2023, https://www.gov.uk/government/news/storm-overflows-monitoring-hits-100-target (accessed 23 May 2025).

88 Environment Agency, 'Environment Agency publishes storm overflow spill data for 2023', 27 March 2024, https://www.gov.uk/government/news/environment-agency-publishes-storm-overflow-spill-data-for-2023 (accessed 23 May 2025).

89 House of Commons Environmental Audit Committee, *Oral Evidence: the Environmental Protection Work of the Environment Agency*, Q.21.

Notes

90 House of Commons Environmental Audit Committee, *Oral Evidence: the Environmental Protection Work of the Environment Agency*, Q.12.

91 Stallard et al., 'Water firms illegally spilled sewage on dry days'.

92 House of Lords Industry and Regulators Committee, *The Affluent and the Effluent*, p. 3.

93 Daniel Capurro, 'Sewage plant inspection targets have been dropped, admits the Environment Agency', *The i Paper*, 7 March 2023, https://inews.co.uk/news/environment-agency-admits-it-has-no-target-for-inspecting-sewage-works-despite-visiting-just-6-in-a-year-2193566 (accessed 23 May 2025).

94 House of Commons Environmental Audit Committee, *Oral Evidence: the Environmental Protection Work of the Environment Agency*, Q.6.

95 Steve Barclay, 'Water company inspections update', statement from Defra, 20 February 2024, https://questions-statements.parliament.uk/written-statements/detail/2024-02-20/HCWS268 (accessed 23 May 2025).

96 NAO, *Achieving the Government's Long-Term Environmental Goals*, 2019–21 HC 958 (2020a), https://www.nao.org.uk/wp-content/uploads/2022/08/Achieving-governments-longterm-environmental-goals.pdf (accessed 23 May 2025).

97 NAO, *Regulating to Achieve Environmental Outcomes*, 2022-23 HC 1283 (2023), https://www.nao.org.uk/wp-content/uploads/2023/04/regulating-to-achieve-environmental-outcomes.pdf (accessed 23 May 2025).

98 Thames Water, 'Section 4: current and future water supply – April 2020', *Final Water Resources Management Plan 2019* (2020), https://www.thameswater.co.uk/media-library/home/about-us/regulation/water-resources/technical-report/current-and-future-water-supply.pdf (accessed 23 May 2025).

99 NAO, *Water Supply and Demand Management*, HC 107 Session 2019-20 (2020b), p. 10, https://www.nao.org.uk/wp-content/uploads/2020/03/Water-supply-and-demand-management.pdf (accessed 23 May 2025).

100 British Geological Survey, 'Groundwater resources in the UK' (2024), https://www.bgs.ac.uk/geology-projects/groundwater-research/groundwater-resources-in-the-uk/ (accessed 23 May 2025).

Notes

101 Cambridge Water (undated), 'Where our water comes from', https://www.cambridge-water.co.uk/environment/managing-water-resources/where-our-water-comes-from/ (accessed 23 May 2025).
102 Defra, *Our Integrated Plan for Delivering Clean and Plentiful Water*, p. 66.
103 South East Rivers Trust (undated), 'Water, water everywhere ... or is it?', https://www.southeastriverstrust.org/water-water-everywhere-or-is-it/ (accessed 23 May 2025).
104 NAO, *Water Supply and Demand Management*, p. 15.
105 John Lawson, *A%R. Review of Abstraction as a % of Recharge in Chalk Streams*, independent report for the CaBA chalk stream restoration group (2021), p. 10, https://www.wildtrout.org/assets/reports/AR-report-27.1.22-web.pdf (accessed 23 May 2025).
106 Office for Environmental Protection, *A Review of Implementation of the Water Framework Directive Regulations and River Basin Management Planning in England* (2024), p. 4, https://www.theoep.org.uk/sites/default/files/reports-files/A%20review%20of%20the%20implementation%20of%20River%20Basin%20Management%20Planning%20in%20England_Accessible.pdf (accessed 23 May 2025).
107 Parliamentary Office of Science & Technology, *Reform of Freshwater Abstraction* (2017), POSTNOTE no.546, Houses of Parliament, https://researchbriefings.files.parliament.uk/documents/POST-PN-0546/POST-PN-0546.pdf (accessed 23 May 2025).
108 Environment Agency, *A Summary of England's Revised Draft Regional and Water Resources Management Plans* (2024), section 3.2, https://www.gov.uk/government/publications/a-review-of-englands-draft-regional-and-water-resources-management-plans/a-summary-of-englands-draft-regional-and-water-resources-management-plans#how-much-water-do-we-need (accessed 23 May 2025).
109 Environment Agency, *A Summary of England's Revised Draft Regional and Water Resources Management Plans*, table 3.
110 Environment Agency, *A Summary of England's Revised Draft Regional and Water Resources Management Plans*, figure 2.
111 House of Commons Environmental Audit Committee, *Oral Evidence: the Environmental Protection Work of the Environment Agency*, p. 22.
112 Environment Agency, *A Summary of England's Revised Draft Regional and Water Resources Management Plans*, section 4.2

Notes

113 House of Commons Environment, Food and Rural Affairs Committee, *Regulation of the Water Industry* 9 October 2018, para 10, https://publications.parliament.uk/pa/cm201719/cmselect/cmenvfru/1041/104104.htm#footnote-183 (accessed 23 May 2025).

114 Ofwat and Environment Agency, *The Case for Change – Reforming Water Abstraction Management in England* (2015), https://www.ofwat.gov.uk/wp-content/uploads/2015/11/pap_pos20111205abstraction.pdf (accessed 23 May 2025).

115 Environment Agency, 'Comply with your water abstraction or impounding licence', 27 March 2018, https://www.gov.uk/guidance/comply-with-your-water-abstraction-or-impounding-licence (accessed 23 May 2025).

116 Amanda Jasi, 'United Utilities fined £800,000 for illegally removing water from the environment', *The Chemical Engineer*, 1 September 2023, https://www.thechemicalengineer.com/news/united-utilities-fined-800-000-for-illegally-removing-water-from-the-environment/ (accessed 23 May 2025).

117 NAO, *Water Supply and Demand Management*, p. 29.

118 Defra, 'Water abstraction statistics: England, 2000 to 2018', 4 January 2023, https://www.gov.uk/government/statistics/water-abstraction-estimates/water-abstraction-statistics-england-2000-to-2018 (accessed 23 May 2025).

119 NAO, *Water Supply and Demand Management*, p. 7.

120 NFU, 'Future changes to abstraction licences – what you need to know', *NFU online*, 25 October 2024, https://www.nfuonline.com/updates-and-information/future-changes-to-abstraction-licences-what-you-need-to-know/ (accessed 23 May 2025).

121 NFU, 'Future changes to abstraction licences – what you need to know'.

122 Defra, *Abstraction Reform Report: Progress Made in Reforming the Arrangements for Managing Water Abstraction in England* (2019), https://assets.publishing.service.gov.uk/government/uploads/system/uploads/attachment_data/file/914427/abstraction-reform-report.pdf (accessed 23 May 2025).

123 Pete Fox, 'Protecting our precious chalk streams', Environment Agency blog, 2 October 2019, https://environmentagency.blog.gov.uk/2019/10/02/protecting-our-precious-chalk-streams/ (accessed 23 May 2025).

Notes

124 Justin Neal, 'Yorkshire anglers win key legal battle and set a precedent for restoring England's rivers and streams', WildFish, 3 April 2025, https://wildfish.org/latest-news/yorkshire-anglers-win-key-legal-battle-and-set-a-precedent-for-restoring-englands-rivers-and-streams/ (accessed 23 May 2025).

125 Rose O'Neill and Kathy Hughes, *The State of England's Chalk Streams*, WWF (2014), https://assets.wwf.org.uk/downloads/wwf_chalkstreamreport_jan15_forweb.pdf?_ga=1.44823268.1991529649.1444910634 (accessed 23 May 2025).

126 Thames Water (undated), 'Teddington Direct River Abstraction (TDRA)', project update, https://thames-sro.co.uk/projects/tdra/ (accessed 23 May 2025).

127 Save Our Lands and River: STOP Thames Water TDRA, https://saveourlandsandriver.org.uk (accessed 23 May 2025).

Chapter 6

1 Foundational Economy Collective (undated), 'What is the foundational economy?', https://foundationaleconomy.com/introduction/ (accessed 23 May 2025).

2 Amartya Sen, *Development as Freedom* (Oxford University Press, 1999).

3 Luca Calafati, Julie Froud, Colin Haslam, Sukhdev Johal and Karel Williams, *When Nothing Works* (Manchester University Press, 2023), pp. 10–16.

4 Anne Lacaton and Jean-Philippe Vassal, *Freedom of Use* (Sternberg Press, 2015).

5 David Edgerton, *The Shock of the Old* (Profile Books, 2006).

6 Calafati et al., *When Nothing Works*, pp. 222–3.

7 Rebecca Speare-Cole, 'Reforms to unlock biggest private investment in water sector's history – Reed', *The Standard*, 5 September 2024, https://www.standard.co.uk/news/politics/water-labour-environment-agency-putney-london-b1180334.html (accessed 23 May 2025).

8 Will Kenton, 'Zombies; financial term for distressed companies', *Investopedia*, 31 August 2021, https://www.investopedia.com/terms/z/zombies.asp (accessed 23 May 2025).

Notes

9 Bernard Condon, 'Zombies: ranks of world's most debt-hobbled companies are soaring, and not all will survive', *AP News*, 7 June 2024, https://apnews.com/article/zombie-business-corporate-debt-investing-interest-borrowing-52bd9ebbe1dd98983d39fe7d14e3c7fd (accessed 23 May 2025).

10 Robert Smith, Gill Plimmer and Josephine Cumbo, 'Thames Water's credit rating slashed to "junk"', *Financial Times*, 24 July 2024, https://www.ft.com/content/7d79cf75-8f67-4f27-a6a1-507640b77f7e (accessed 23 May 2025).

11 BBC News, 'Southern Water credit rating downgraded to "junk"', 14 November 2024, https://www.bbc.co.uk/news/articles/cj9nrjpjk8jo (accessed 23 May 2025).

12 Euan Healey, 'Thames Water and Altice set to push European high yield default rate to highest since 2008', *Financial Times*, 3 February 2025, https://www.ft.com/content/84be7389-cca7-4865-9f28-ee54edfbd28d (accessed 23 May 2025).

13 Robert Smith, 'Thames Water £3bn rescue loan gets High Court approval', *Financial Times*, 18 February 2025, https://www.ft.com/content/b6bb1eb3-f4f3-4a8b-94fa-00ebf7f25d28 (accessed 23 May 2025).

14 Smith, 'Thames Water £3bn rescue loan gets High Court approval'.

15 Source: Company report and accounts, various years.

16 Ofwat, *Our Final Determinations for the 2024 Price Review: Sector Summary* (2024), p. 6, https://www.ofwat.gov.uk/wp-content/uploads/2024/12/PR24-FD-sector-summary.pdf (accessed 23 May 2025).

17 Water and Waste Water Company PR24 draft determination Business Plans, sheet RR19.

18 Ofwat, 'Monitoring financial resilience report 2023–24 charts and underlying data', 21 November 2024, https://www.ofwat.gov.uk/publication/monitoring-financial-resilience-report-2023-24-charts-and-underlying-data/ (accessed 23 May 2025).

19 Ofwat, 'Monitoring financial resilience report 2023–24 charts and underlying data'.

20 Robert Smith, Jim Pickard and Kieran Smith, 'Water bills in England and Wales to rise 36% after Ofwat review', *Financial Times*, 19 December 2024, https://www.ft.com/content/039c2832-c257-4fd9-8220-eacb3c1687ca (accessed 23 May 2025).

Notes

21 Clear Barrett and Gill Plimmer, 'Thames Water customers shocked by "scandalous" bill increases', *Financial Times*, 28 February 2025, https://www.ft.com/content/d514c59f-bcc1-42b7-9f1b-fa517c3d00d6 (accessed 23 May 2025).

22 Kyriakos Petrakos, 'Thames super-sewer adds £160 to water bills while CEO takes home extra £1.7m', *The i Paper*, 30 November 2024, https://inews.co.uk/news/thames-water-increasing-bills-160-chief-exec-1-7m-3405102 (accessed 23 May 2025).

23 Ofwat, *Major Projects Development and Delivery* (2024), pp. 9–10, https://www.ofwat.gov.uk/wp-content/uploads/2024/12/PR24-final-determinations-Major-Projects-development-and-delivery-V2.pdf (accessed 23 May 2025).

24 Ofwat, *Major Projects Development and Delivery*, p. 8.

25 Ofwat, *Major Projects Development and Delivery*, pp. 22–3.

26 Defra, *Call For Evidence. Independent Commission on the Water Sector Regulatory Scheme (2025)*, p. 197, https://consult.defra.gov.uk/independent-water-commission/independent-commission-on-the-water-sector-regulat/supporting_documents/Call%20For%20Evidence%20%20Independent%20Commission%20on%20the%20Water%20Sector%20Regulatory%20System.pdf (accessed 23 May 2025).

27 Dieter Helm, 'More debt, less competition: Ofwat's DPC and SIPR proposals', blog, 22 July 2024, https://dieterhelm.co.uk/regulation-utilities-infrastructure/more-debt-less-competition-ofwats-dpc-and-sipr-proposals/ (accessed 23 May 2025).

28 Ofwat, *Major Projects Development and Delivery*. There are 31 projects listed by Ofwat and we list 27. The criteria for inclusion are 1) they must be the responsibility of at least one of the ten water companies that we are analysing; 2) we can assign the disclosed individual projects to each water/wastewater company; and 3) we can align them to the enhancement capital expenditure disclosed in the PR24 business plans of each company.

29 Ofwat, *Major Projects Development and Delivery*, p. 2.

30 Thames Tideway has had this status since 2015. Tideway, 'Annual Report 2020/21' (2021), p. 2, https://www.tideway.london/media/5072/tideway-annual-report-2020-2021.pdf (accessed 23 May 2025).

Notes

31 National Audit Office, *PFI and PF2,* HC 718, session 2017-2019 (2018), https://www.nao.org.uk/wp-content/uploads/2018/01/PFI-and-PF2.pdf (accessed 23 May 2025).

32 Ofwat, *Major Projects Development and Delivery: PR24 Draft Deliberations* (2024), p. 1, https://www.ofwat.gov.uk/wp-content/uploads/2024/07/PR24-draft-determinations-Major-projects-development-and-delivery-1.pdf (accessed 23 May 2025).

33 Benjamin Barber, *Strong Democracy* (University of California Press, 1984).

34 Michael Moran, *Politics and Governance in the UK* (Bloomsbury, 2015), p. 7.

35 Ed Atkins, 'Building a dam, constructing a nation: the "drowning" of Capel Celyn', *Journal of Historical Sociology*, 31.4 (2018), 455–68, https://onlinelibrary.wiley.com/doi/10.1111/johs.12186 (accessed 23 May 2025).

36 Save Windermere (undated), 'A historic victory for Windermere – but our fight isn't over', https://www.savewindermere.com/sewage-free-windermere (accessed 23 May 2025).

37 Source: M. J. Durant and C. J. Counsell, 'Inventory of reservoirs amounting to 90% of total UK storage', NERC Environmental Information Data Centre (2018), https://catalogue.ceh.ac.uk/documents/f5a7d56c-cea0-4f00-b159-c3788a3b2b38 (accessed 23 May 2025).

38 Stephen V. Ward, 'An essay in civilisation', *Angles*, 15 (2022), 4–6.

39 Ben Brown, 'From "no-go" zones to bohemian utopia: what happened to Hulme's notorious "crescents"?', MCR Finest (2022), https://www.manchestersfinest.com/articles/hulmes-notorious-crescents/ (accessed 23 May 2025).

40 Pippa Crerar, 'Work on up to 12 new towns in England to begin by next election, says government', *The Guardian*, 12 February 2025, https://www.theguardian.com/society/2025/feb/12/up-to-12-new-towns-will-be-under-construction-in-england-by-next-election-says-starmer (accessed 23 May 2025).

41 Aeron Davis, *Bankruptcy, Bubbles and Bail Outs* (Manchester University Press, 2022).

42 Will Dunn, 'Revealed: the 103 professional lobbyists standing in the 2024 general election', *New Statesman*, 3 July 2024, https://

Notes

www.newstatesman.com/politics/2024/07/revealed-the-103-professional-lobbyists-hoping-to-become-mps (accessed 23 May 2025).

43 More in Common, 'Climate and energy at the General Election', 10 July 2024, https://www.moreincommon.org.uk/latest-insights/climate-and-energy-at-the-general-election/ (accessed 23 May 2025).

44 Lena Swedlow and Clive Lewis, *Our Water Our Way*, Compass (2025), p. 5, https://www.compassonline.org.uk/publications/our-water-our-way/ (accessed 23 May 2025).

45 Compass, 'Thames Water emergency board launched!', 11 March 2025, https://www.compassonline.org.uk/thames-water-emergency-board-launched/ (accessed 23 May 2025).

46 Labour Party, *Democratic Public Ownership*, Labour Party Consultation Paper (2018), https://labour.org.uk/wp-content/uploads/2018/09/Democratic-public-ownership-consulation.pdf (accessed 23 May 2025).

47 Stephen Elstub, David M. Farrell, Jayne Carrick and Patricia Mockler, *Evaluation of Climate Assembly UK*, Newcastle University (2021), https://www.parliament.uk/globalassets/documents/get-involved2/climate-assembly-uk/evaluation-of-climate-assembly-uk.pdf (accessed 23 May 2025).

48 Climate Assembly UK, *The Path to Net Zero: Climate Assembly UK Full Report* (2020), pp. 16–17, https://www.climateassembly.uk/report/read/final-report.pdf (accessed 23 May 2025).

49 Business, Energy and Industrial Strategy Committee, *Climate Assembly UK: where are we now?*, second report of session 2021–22, HC 546 (2020), https://committees.parliament.uk/publications/6617/documents/71408/default/ (accessed 23 May 2025).

50 Michela Palese, 'The Irish abortion referendum: how a Citizens' Assembly helped to break years of political deadlock', Electoral Reform Society, 29 May 2018, https://electoral-reform.org.uk/the-irish-abortion-referendum-how-a-citizens-assembly-helped-to-break-years-of-political-deadlock/ (accessed 23 May 2025).

51 March for Clean Water, https://marchforcleanwater.org/supporters (accessed 23 May 2025).

52 The End Sewage Pollution Manifesto, https://www.sas.org.uk/wp-content/uploads/2023/09/End-Sewage-Pollution-Manifesto-Final.pdf (accessed 23 May 2025).

Notes

53 Donatella Della Porta, 'Building bridges: social movements and civil society in times of crisis', *Voluntas*, 31 (2020), 938–48.
54 Jerry van den Berge, Jeroen Vos and Rutgerd Boelens, 'Water justice and Europe's Right2Water movement', *International Journal of Water Resources Development*, 38.1 (2022), 173–91, DOI: 10.1080/07900627.2021.1898347; Transnational Institute, *Here to Stay: Water Remunicipalisation as a Global Trend* (2014), https://www.tni.org/en/publication/here-to-stay-water-remunicipalisation-as-a-global-trend (accessed 23 May 2025).
55 European Water Movement (undated), 'The Naples Manifesto', https://europeanwater.org/about-the-european-water-movement/naples-manifesto (accessed 23 May 2025).
56 European Water Movement (undated), 'The European Water Movement in the Face of the European Economic and Social Committees' Blue Deal Initiative', https://europeanwater.org/images/pdf/EWM-PositionBlueDeal_EN.pdf (accessed 23 May 2025).
57 See, for example, the European Citizens' Initiative 'Right to Water' campaign, https://citizens-initiative.europa.eu/news/follow-first-ever-successful-eci-right-water_en (accessed 23 May 2025).
58 Windrush Against Sewage Pollution (undated), 'Investigating the health of our rivers', https://www.windrushwasp.org/data-analysis (accessed 23 May 2025).
59 Watershed (undated), 'Recent Investigations', https://watershedinvestigations.com (accessed 23 May 2025).
60 Watershed (undated), 'The Watershed pollution map', https://watershedinvestigations.com/map-whats-polluting-your-local-river-lake-or-coast/ (accessed 23 May 2025); Rachel Salvidge and Leana Hosea, 'Revealed: drinking water sources in England polluted with forever chemicals', *The Guardian*, 16 January 2025, https://www.theguardian.com/environment/2025/jan/16/the-forever-chemical-hotspots-polluting-england-drinking-water-sources (accessed 23 May 2025).
61 https://waterwatchuk.org/about/ (accessed 23 May 2025).
62 https://extinctionrebellion.uk/act-now/campaigns/dont-pay-for-dirty-water/ (accessed 23 May 2025).
63 Jayne Mann, 'Feargal Sharkey demands immediate action to restore rivers back to sustainable flows during the Rivers Trust Autumn Conference',

Notes

The Rivers Trust, 22 November 2018, https://theriverstrust.org/about-us/news/feargal-sharkey-demands-immediate-action-to-restore-rivers-back-to-sustainable-flows-during-the-rivers-trust-autumn-conference-2018 (accessed 23 May 2025).

64 Surfers Against Sewage (undated), 'About us', https://www.sas.org.uk/about-us/ (accessed 23 May 2025).

65 Gill Plimmer, 'Water companies face legal challenges after landmark UK pollution ruling', *Financial Times*, 2 July 2024, https://www.ft.com/content/9e7e840e-6d16-47ee-8ac1-9586a3de1495 (accessed 23 May 2025).

66 River Action, 'Campaign group to appeal legal challenge against the Environment Agency & prepares for further legal action to protect the Wye', 18 June 2024, https://riveractionuk.com/news/campaign-group-to-appeal-legal-challenge-against-the-environment-agency-prepares-for-further-legal-action-to-protect-the-wye/ (accessed 23 May 2025).

67 Surfers Against Sewage (undated), 'Ban the Bailouts', https://www.sas.org.uk/water-quality/our-water-quality-campaigns/ban-the-bailouts/#newmode-embed-52564-69115 (accessed 23 May 2025).

68 The End Sewage Pollution Manifesto, https://www.sas.org.uk/wp-content/uploads/2023/09/End-Sewage-Pollution-Manifesto-Final.pdf (accessed 23 May 2025).

69 March for Clean Water (undated), 'The coalition', https://marchforcleanwater.org/supporters (accessed 23 May 2025).

70 The Rivers Trust, 'The Rivers Trust response: Environmental Audit Committee's Water Quality and Water Infrastructure: follow-up inquiry', written evidence (2024), p. 5, https://committees.parliament.uk/writtenevidence/130196/pdf/ (accessed 23 May 2025).

71 SSWAN (undated), 'A new approach', https://www.sswan.co.uk/#group-section-A-new-approach-sZtRxnutpg (accessed 23 May 2025).

72 SSWAN (undated), *Sustainable Solutions for Water and Nature. Discussion Paper*, p. 1, https://www.wessexwater.co.uk/media/4popagnb/sswan-discussion-paper.pdf (accessed 23 May 2025).

73 Civil Liberties Union for Europe, *Liberties Rule of Law Report* (2025), https://www.liberties.eu/f/vdxw3e (accessed 23 May 2025).

Index

abstraction (of water) 47, 61, 207–8, 213–19
affordability 9, 88, 143–50, 160–1
  see also water justice
Anglers Trust 269
Anglian Water 85, 91
aquifer 214–15
asset intensity 13–14, 74–7, 79–81, 110, 194–5
  see also business model of water companies

Bayliss, Kate 2, 70–1, 73, 93
Bazalgette, Joseph 43, 45–6
business customers 101–2, 103–4
  charges 102
  competition 102–4
business model of water companies 13, 66–7, 89, 93, 100–1, 142, 170, 228, 241
  activity characteristics 13–14, 67–8, 74–6
Byatt, Ian 196–7

Capel Celyn 22, 43–4, 49, 253
catchment-based approaches 34, 135–7, 185, 197, 218, 233–5, 237, 249, 263, 270

challenging power 25–6
citizens' assemblies 26, 262, 263
citizenship 28–9, 225, 235, 252–3, 272–3
civil society 3–4, 27, 38, 117, 143, 183–4, 190, 207, 220, 235, 264–5, 272
  power struggle 252, 254, 266–7
  see also social movement in water
climate change 5, 8, 126–7, 128–9, 260
  adaptation to 5, 20, 39, 126–7, 136, 226–7, 235, 260
  impact on water supply 130–3, 137–8
  mitigation 29, 39, 51, 53, 56, 126
  water management 234–5
  see also Environment Agency; flooding; precipitation
combined sewerage system 19, 36–7, 43, 113
Compass 2, 5, 262
Consumer Council for Water (CCW) 145, 147, 148, 160

Index

cost-of-living crisis 144, 146–7, 172–3, 260
Costa Beck 219
Cunliffe Independent Commission on Water 5, 25, 38, 104, 187, 190–1, 245, 261

Davis, Aeron 260
debt (in the water companies) 2, 14, 25, 66–8, 70–1, 88, 97–9, 193, 198, 201–2, 240–2, 243
　PR24 plans 242–3
decoupling of emissions 54–5
Defra 10, 18, 125, 130, 144, 180, 189, 204–5, 212–13, 249
　and abstraction 214, 217, 218–19
　Plan for Water 135–7
Della Porta, Donatella 264
democracy (in water) 26, 28, 261–3, 272
Direct Procurement for Customers (DPC) 246–7, 248, 250–1
　see also projectification
Dŵr Cymru 14, 92–3, 97, 98–100, 145, 161, 231

echo chamber politics 181–2, 183, 186–7, 189
　see also issue management
Elan Valley 22, 42–3, 48–9
End Sewage Pollution Manifesto 6, 264, 269–70
Environment Act 2021 218
Environment Agency 24, 108, 125, 131, 132, 185, 204, 205–9, 211, 218, 219, 238, 267
　abstraction 215, 216
　and flood risk 61, 135–6
　monitoring and inspections 3, 122–3, 208–9, 211–12

storm overflows 120–1, 210–11
　National Framework for Water Resources 131–2
　Regional and Water Resources Management Plans 132–3
European Union Water Framework Directive 123, 207, 215
European water movement 265–7
Extinction Rebellion 268

financial engineering 2, 14, 25, 94, 98, 100, 197–9, 201–3, 232–3, 248, 258
　see also projectification; zombie companies
financial extraction 2, 4, 13, 68, 69–72, 95, 199, 202, 241
financial unsustainability 14, 79–80, 192
Fish Legal 219, 268
flooding 19, 56, 61, 257
　increased risk of 136, 226, 234
foundational economy 2, 7–8, 28–9, 225
foundational water management 8–9, 224–7, 228, 238, 251–2, 261
　technical reforms 229–36
Fressoz, Jean-Baptiste 53

General Election 2024 (and water reform) 186–8
Gove, Michael 2, 72, 73, 201

Hall, David 2, 70, 71, 201
Helm, Dieter 36, 70–1, 194, 244
household water bills 9, 26, 82–3, 85, 117, 147, 148–50
　see also affordability; progressive charging; regressive charging; social tariffs

327

Index

intercept sewers 19, 40–1, 43, 45–6
Intergovernmental Panel on Climate Change (IPCC) 52–3
investment 13, 18, 66–8, 71, 76, 110–14, 117–18, 156–66, 191, 194–5
   need for investment 51, 61, 83, 136, 185, 187, 190, 197, 266
   renewal rate 111–13, 209, 133
   under-investment 36–8, 44, 72, 77, 84, 88, 101, 107–8, 156, 191
   see also PR19; PR24; projectification
issue management 138, 182–4, 186, 190–1, 212–13, 257

Kemble Water Holdings 90, 200
   see also Thames Water

labour costs 67, 77–9
leakage 37, 133–4
   reduction 118, 121, 132, 133–4, 135, 216
   see also PR24
Lewis, Clive 262
Lobina, Emanuele 70

Macquarie 14, 24, 68, 70, 90–1, 97–8, 198, 203, 240
   see also financial engineering
March for Clean Water 6, 264, 269
meters 15, 132–3, 134, 144, 148, 149, 173

narratives 4, 17–18, 19, 117, 125, 183
   of the water industry 18, 20, 68, 71, 125
   see also Water UK

National Audit Office 104, 151, 212, 247
   abstraction 214, 218
   storm overflows 210
National Grid 74, 76, 80–1
National Infrastructure Commission 4, 130–1, 184, 185
nature-based solutions 126, 127, 136–7, 187, 197, 233–4, 236–7, 249, 270, 271
net zero 20, 55–6, 257, 260
Northumbrian Water 90, 145

Office for Environmental Protection 184, 185–6, 215, 258
Ofwat 24, 82–5, 191–3, 196, 197, 201–3, 205, 209
   asset health 111, 115–16
   capital expenditure 112–13, 114, 115
   enhancement expenditure 66–7, 69, 77, 113–14, 245
   operating expenditure (OPEX) 110–11
   performance review 115–16
   price-setting process 193–5, 203
   total expenditure (TOTEX) 193, 248
   water poverty 146, 147
   see also PR19; PR24
operator self-monitoring system 206–7
ownership of water companies 90–2, 95–7
   not-for-profit 93
   public listed companies 92, 95–7
   see also public ownership

Index

Plimmer, Gill 72
power 11, 20–2, 33, 194, 197, 229
   democracy 25–6, 38, 252–3, 261–5, 272–3
   elite dysfunction 181, 249, 257–9
   foundational provision 254–6
   power configuration 12, 23–4, 26–8, 50, 230, 252, 254–5, 256, 260
   technocracy 257–8
   see also social movement for water
PR19 73, 82–3, 88, 112, 113–14, 194, 201–2, 243
   see also Ofwat
PR24 25, 82–4, 88, 112–13, 114, 118, 121, 126, 148, 156, 161–2, 194, 242–3
   privately financed projects 245, 250
   see also Ofwat
precipitation 60, 61
   past trends 57–9, 59–60
   projections 128–9, 130–1
   see also climate change
Princess Alice disaster 19, 46
privately financed water projects 108, 121, 156, 210, 245–8, 256–7
   not in the public interest 248–51
   see also projectification
privatisation of water in England and Wales 2, 23, 34–5, 66, 70, 73, 82, 88, 89–90, 102, 143, 192, 233
   public opposition to 5, 27, 69, 183–4
productivity in the water industry 78–9
progressive charging 16, 164, 168, 169–71, 174–6

flat rate charging 165–9
   reform 229–31
projectification 25, 210, 228, 239, 244–5, 248, 252, 258, 259
public ownership 5–6, 16–17, 92, 93–4, 186–7
   Labour Party 2018 ownership plan 14–15, 94, 262
   reform 25, 100, 231–3, 262–3

Reed, Steve 18, 23–4, 38, 83, 189–90
regressive charging 15, 82, 142, 150–1, 153–5, 231, 270
reservoirs 21, 44, 237, 246, 259
   history of construction 21–2, 41, 42, 43–4, 47, 48–50, 255–6
revenue constraint 13, 81–2, 85–7, 154–6, 187, 230
River Action 3, 27, 47, 124, 191, 268–9
Rivers Trust 4, 38, 268, 269, 270
road runoffs 3, 122, 124–5, 137, 212

sanitation, history of 19, 40–1, 42, 44–5, 51, 61, 228
Save Windermere 253
Scottish Water 92–3
sewage discharges 14, 19, 51, 118, 122, 123, 124, 204, 210, 214
sewage treatment 41–2, 44, 46–7
Shaoul, Jean 70, 79
Sharkey, Feargal 190–1, 268
social movement for water 26, 27, 261, 264–7, 272–3
social tariff 15, 16, 142, 144, 145–6, 148, 159, 161–2, 163, 187, 239
   national social tariff 85, 142, 148, 151, 159–60, 236
   WaterSure 144, 145, 147

329

Index

South West Water 90, 161
Southern Water 90, 91, 162, 203, 209, 211, 240–1
Specified Infrastructure Projects Regulations (SIPR) 246–7, 248–51, 258
  see also projectification
storm overflows 2, 17, 18, 19, 45, 51, 66, 72, 117–18, 120–1, 186, 188, 208, 210, 257, 272
  data and monitoring 108, 210–11
  dry spills 2, 206, 211
  see also Water UK *National Storm Overflows Plan*
Surfers Against Sewage 47, 268, 269
Sustainable Development Goals (SDGs) 33, 142
Sustainable Solutions for Water and Nature (SSWAN) 27, 270–1

temperature change 4, 20, 52–3, 56
  past trends 56, 57–9
  projections 128–31
Thames Water 66, 90–1, 97–8, 114, 190, 203, 209, 240–1, 246, 259, 262
  abstraction 47, 213–14, 215, 220
  corporate structure 199–200
  debt 14, 70, 98–9, 177
  infrastructure condition 108–10
  sanitation crisis 19, 45, 46, 206, 211, 213
  Thames Tideway 156, 239, 245, 247, 250
  water bills 85, 161–2, 244–5
  see also Macquarie
Thirlmere (reservoir) 40, 42, 44, 48
trade associations 17–18, 116–17, 118
  see also Water UK

United Utilities 3, 85, 91, 114, 161–2, 206, 207, 217

water bills 82–3, 154–6
water consumption 3, 44, 132–5
water cycle 38–9
water justice 16, 160–4, 170–2, 176–7
water management 28, 51, 183, 224–5, 226–7, 228–9, 234–5, 236, 249, 250, 257, 262, 264–5, 272
  see also foundational water management
water poverty 16, 141–3, 144–5, 146–7, 148
  disconnections 143
water quality 38, 47, 83, 122–5, 137, 184–5, 205, 207–8, 211–12, 214–15, 268
Water Resources Management Plans 130, 132, 216, 247
Water Special Measures Bill 2024 18, 189–91, 269
Water UK 6, 18, 83–4, 117, 142, 194, 207
  National Storm Overflows Plan 118–23, 125
  water poverty 146, 148
Watershed Investigations 267–8
Waterwatch 268
Watson, Charles 191
Wessex Water 90, 91, 95, 211, 271
Windrush Against Sewage Pollution (WASP) 3, 267, 269
Wye Valley 3, 123–4

Yorkshire Water 91, 201, 209

zombie companies 25, 239–43

330

EU authorised representative for GPSR:
Easy Access System Europe, Mustamäe tee 50,
10621 Tallinn, Estonia
gpsr.requests@easproject.com